Modern Electrosynthetic Methods in Organic Chemistry

New Directions in Organic & Biological Chemistry

Series Editor:
Philip Page

PUBLISHED TITLES

C-Glycoside Synthesis
Maarten Postema

Capillary Electrophoresis: Theory and Practice
Patrick Camilleri

Carbocation Chemistry: Applications in Organic Synthesis
Jie Jack Li

Chemical Approaches to the Synthesis of Peptides and Proteins
Paul Lloyd-Williams, Fernando Albericio, and Ernest Giralt

Chiral Sulfur Reagents
M. Mikolajczyk, J. Drabowicz, and P. Kielbasinski

Chirality and the Biological Activity of Drugs
Roger J. Crossley

Concerted Organic and Bio-organic Mechanisms
Andrew Williams

Cyclization Reactions
C. Thebtaranonth and Y. Thebtaranonth

Dianion Chemistry in Organic Synthesis
Charles M. Thompson

Mannich Bases: Chemistry and Uses
Maurilio Tramontini and Luigi Angiolini

Modern Electrosynthetic Methods in Organic Chemistry
Frank Marken & Mahito Atobe

Modern NMR Techniques for Synthetic Chemistry
Julie Fisher

Organozinc Reagents in Organic Synthesis
Ender Erdik

Modern Electrosynthetic Methods in Organic Chemistry

Edited by
Professors Frank Marken and Mahito Atobe

CRC Press
Taylor & Francis Group
Boca Raton London New York

CRC Press is an imprint of the
Taylor & Francis Group, an **informa** business

CRC Press
Taylor & Francis Group
6000 Broken Sound Parkway NW, Suite 300
Boca Raton, FL 33487-2742

First issued in paperback 2020

ISBN-13: 978-1-4822-4916-3 (hbk)
ISBN-13: 978-0-367-73287-5 (pbk)

Visit the Taylor & Francis Web site at
http://www.taylorandfrancis.com

and the CRC Press Web site at
http://www.crcpress.com

Contents

Preface...vii
Editors...ix
Contributors xi

Chapter 1 Introduction to Practical Methods in Electroorganic Syntheses..........1

      James E. Taylor, M. Anbu Kulandainathan,
      Christopher E. Hotchen, and Frank Marken

Chapter 2 Cogeneration Electrosynthesis 45

      Xiangming Gao, Adrian C. Fisher, Nathan Lawrence,
      and Peng Song

Chapter 3 Photoelectrochemically Driven Electrosynthesis..............................67

      Hisashi Shimakoshi and Yoshio Hisaeda

Chapter 4 Ultrasound Activation in Electrosynthesis..81

      Mahito Atobe and Frank Marken

Chapter 5 Novel Electrode Materials in Electroorganic Synthesis....................97

      Christoph Gütz, Sebastian Herold, and Siegfried R. Waldvogel

Chapter 6 Electrosynthesis of Functional Polymer Materials 127

      Shinsuke Inagi and Naoki Shida

Chapter 7 Biphasic and Emulsion Electroorganic Syntheses 149

      Sunyhik Ahn and Frank Marken

Chapter 8 Solid-to-Solid Transformations in Organic Electrosynthesis 167

      Antonio Doménech-Carbó

Index..181

Preface

Electro-organic synthesis enters a new era due to the ongoing transformation of synthetic chemistry from fossil-based to renewable-based and due to the desire to automate and control synthetic reactions. It has long been realized that electrochemical processes offer excellent atom economy and low-waste synthetic routes, and occasionally elegant reaction pathways with readily available radical intermediates. But significant obstacles to widespread use remain, for example, in the perceived need for excess supporting electrolyte and the resulting complexity of workup, separation, and reuse of electrolyte. In fact, many electro-organic reactions should work well even in the absence of supporting electrolyte. For example, "self-supported" reactions exist and "bead-supported" electrolytes have been introduced by Breinbauer and coworkers [1] and by Fuchigami and coworkers [2]. There is still a need to develop/commercialize and explore new flow reactor systems for systematic synthetic processes with multiple stages or with coupled or paired reactions. Currently, electrosynthetic instrumentation is virtually absent from organic synthetic laboratories, and therefore, electrosynthetic routes remain very much underdeveloped and underexploited at all levels. Recently, Baran and coworkers highlighted the promise as well as the need for new easily accessible electrosynthesizers [3].

In this compilation of chapters on modern electrosynthetic methods, the main theme is linked to new types of methodology and reaction activation that can be applied in electrosynthesis. This includes an introduction to experimentally simple bulk and flow synthesis, but also information about photo- and ultrasound-activated processes. In Chapter 1, James E. Taylor and coworkers summarize some of the recent developments, and they indicate where innovative electro-organic processes are employed to induce chirality, to provide experimentally simple laboratory tools, and to allow elegant solutions to synthesis steps via flow reactors and paired electrolysis processes.

Xiangming Gao and coworkers explain how cogeneration electrosynthesis can be used to bridge the technology gap between flow cell reactors and driven fuel cell systems. There is an obvious link between energy consumption and energy generation in these types of reactor cells, and there are opportunities for industrial synthetic processes that intelligently couple anodic and cathodic reactions. Shimakoshi and Hisaeda describe recent developments in photo-driven electrosynthesis, which is closely linked to natural photosynthesis. Light energy (as a natural resource) can be employed in the form of electricity, but also directly employed in the activation of reaction intermediates and in the nonthermal control of reaction pathways. Chapter 4 explains how power ultrasound can be employed to affect reaction pathways or reaction speed in electro-organic reactions. Sonoelectrosynthesis (although requiring additional instrumentation) can be highly effective in processes where surface need to be activated.

Material aspects are crucially important for electrode systems and in electrocatalysis and occasionally new materials completely change the way/pathway by which electron transfer occurs. An excellent example for this is the introduction of chemically robust and versatile boron-doped diamond electrodes to electro-organic synthesis, in particular for anodic processes such as hydroxylations. In Chapter 5, Waldvogel and coworkers describe the importance of new materials, and they summarize the latest achievements based on novel alloys that outperform traditional cathode electrode systems.

Multi-phase approaches are highly interesting and both emulsion methods as well as solid dispersion methods are applicable in organic electrochemical transformations. Polymer electrosynthesis produces a solid product that needs to be dispersed while reaction conditions need to be found to provide uniform and reproducible polymer materials. Inagi and Shida provide information about electro-polymerization processes in Chapter 6. Ahn and coworkers then describe modern approaches to emulsion electrosynthesis and biphasic reaction conditions. This includes traditional emulsion conditions but also attempts to engineer three-phase junction interfaces, where rate limitations imposed by poor solubility can be overcome. Finally, in Chapter 8, the idea of solid-to-solid state electro-organic transformation is discussed by Antonio Doménech-Carbó. Although a niche area of electro-organic synthesis, this type of transformation (similar to tribo-chemical reactions) promises clean reactions under conditions of mechanical activation and usually under ambient conditions and in aqueous electrolyte media.

We are grateful for the initial encouragement for this volume from the series editor Professor Philip Bulman Page. We indebted to and very thankful for the time spend and the energy contributed by all authors and coauthors as well as very thankful for the support given by the Taylor & Francis team (especially Hilary Lafoe). We thank you for the unwavering trust in our ability to assemble this volume of contributions to the field of modern electro-organic synthesis.

REFERENCES

1. Nad, S.; Breinbauer, R. 2004. Electroorganic synthesis on the solid phase using polymer beads as supports: Record contains structures. *Angew. Chem. Int. Ed.* 43:2297–2299.
2. Tajima, T.; Fuchigami, T. 2005. Development of a novel environmentally friendly electrolytic system by using recyclable solid-supported bases for in situ generation of a supporting electrolyte from methanol as a solvent: Application for anodic methoxylation of organic compounds. *Chemistry* 11:6192–6196.
3. Yan, M.; Kawamata, Y.; Baran, P.S. 2018. Synthetic organic electrochemistry: Calling all engineers. *Angew. Chem. Int. Ed.* 57:4149–4155.

Frank Marken
Department of Chemistry
University of Bath
Bath, UK

Mahito Atobe
Department of Environment and System Sciences
Yokohama National University
Yokohama, Japan

Editors

Dr. Mahito Atobe was appointed to a Professor position in July 2010 at the Graduate School of Environment and Information Sciences, Yokohama National University. His current research focuses on organic electrosynthetic processes and electrochemical polymerization under ultrasonication, electrosynthetic processes in a flow microreactor, and organic electrochemical processes in supercritical fluids (140 publications).

Dr. Frank Marken was appointed to a Senior Lecturer position in September 2004, promoted to Reader (2008), and to Professor (2011) at the Department of Chemistry, University of Bath. His general research interests are centered around the fundamental mechanistic understanding of interfacial processes and the development of novel electrosynthetic technologies based on nanostructures, three-phase junctions, light- and microwave-enhanced processes, paired and self-supported electrosyntheses, and electrocatalytic reactor development (450 publications, h-index 55).

Contributors

Sunyhik Ahn
Department of Chemistry
University of Bath
Bath, United Kingdom

M. Anbu Kulandainathan
Electro Organic Division
Central Electrochemical Research
   Institute
Karaikudi, India

Mahito Atobe
Department of Environment and System
   Sciences
Yokohama National University
Yokohama, Japan

Antonio Doménech-Carbó
Department of Analytical Chemistry
University of Valencia
Valencia, Spain

Adrian C. Fisher
Department of Chemical Engineering
University of Cambridge
Cambridge, United Kingdom

Xiangming Gao
Department of Chemical Engineering
University of Cambridge
Cambridge, United Kingdom

Christoph Gütz
Institut für Organische Chemie,
Johannes Gutenberg-Universität
   Mainz
Mainz, Germany

Sebastian Herold
Institut für Organische Chemie,
Johannes Gutenberg-Universität Mainz
Mainz, Germany

Yoshio Hisaeda
Department of Chemistry and
   Biochemistry
Graduate School of Engineering,
   Kyushu University
Motooka, Japan

Christopher E. Hotchen
Department of Chemistry
University of Bath
Bath, United Kingdom

Shinsuke Inagi
Department of Chemical Science and
   Engineering,
School of Materials and Chemical
   Technology
Tokyo Institute of Technology
Yokohama, Japan
and
Department of Electronic Chemistry
Interdisciplinary Graduate School of
   Science and Engineering
Tokyo Institute of Technology
Yokohama, Japan

Nathan Lawrence
Department of Chemical Engineering
University of Hull
Hull, United Kingdom

Frank Marken
Department of Chemistry
University of Bath
Bath, United Kingdom

Naoki Shida
Department of Electronic Chemistry
Interdisciplinary Graduate School of
   Science and Engineering
Tokyo Institute of Technology
Yokohama, Japan

Hisashi Shimakoshi
Department of Chemistry and
    Biochemistry
Graduate School of Engineering,
    Kyushu University
Motooka, Japan

Peng Song
Beijing Key Laboratory for Green
    Catalysis and Separation
Department of Chemistry and Chemical
    Engineering
Beijing University of Technology
Beijing, P.R. China

James E. Taylor
School of Chemistry
University of St Andrews
Fife, Scotland

Siegfried R. Waldvogel
Institut für Organische Chemie
Johannes Gutenberg-Universität Mainz
Mainz, Germany

1 Introduction to Practical Methods in Electroorganic Syntheses

James E. Taylor
University of St Andrews

M. Anbu Kulandainathan
Central Electrochemical Research Institute

Christopher E. Hotchen and Frank Marken
University of Bath

CONTENTS

1.1 Introduction to Electrosynthetic Methods 1
    1.1.1 Approaching Electroorganic Synthesis 4
    1.1.2 Mass Transport 7
        1.1.2.1 Mass Transport and Computational Mechanistic Tools 9
    1.1.3 Solvents and Electrolytes 10
        1.1.3.1 Organic Solvents 10
        1.1.3.2 Ionic Liquids 11
        1.1.3.3 Polymer Solvents 14
1.2 Methodology for Direct Transformations 15
    1.2.1 Anodic Processes 15
    1.2.2 Cathodic Processes 22
1.3 Methodology for Indirect and Catalytic Transformations 24
    1.3.1 Anodic Processes 24
    1.3.2 Cathodic Processes 30
1.4 Methodology for Paired Transformations 31
1.5 Outlook, Conclusions, and Recommendations 34
References 35

1.1 INTRODUCTION TO ELECTROSYNTHETIC METHODS

In a recent comprehensive review on electrosynthesis with focus on innovative approaches going beyond traditional texts [1–4], Yan, Kawamata, and Baran [5] emphasize the versatility of electroorganic transformations in a diverse range of

1

challenging laboratory syntheses. Similarly, in a recent topical review by Waldvogel and coworkers [6], new directions and opportunities in organic electrosynthesis are laid out. A huge range of innovative processes including anodic oxidations, cathodic reductions, and catalytic and paired electrolyses are covered with focus on recent progress. A forthcoming "renaissance" of electroorganic processes is suggested linked to (A) the inherent "greenness" of using electrons as reagents provided at an appropriately applied energy/potential [7–9] and (B) the expected transition of traditional "fossil chemistry" to "solar chemistry" based predominantly on electrochemically driven processes [10]. This transition will require many new synthetic tools and could include electroorganic syntheses that "couple" into microbial reactions [11] or that are based on electro-biocatalysis [12].

The use of electrochemical methodology has many potential advantages over traditional organic syntheses [9,13]. Processes can be switched on and off or regulated/optimized in a way more similar to biosynthetic reactions. Often reactions are performed under potential control (potentiostatic control) to provide/remove electrons at fixed energy, which helps to avoid undesired side reactions. Sometimes "solvated electrons" emitted from the cathode (e.g., in the Birch reactions [14]) are employed directly in cathodic electrosyntheses [15,16] or, otherwise, redox mediators can be employed to couple heterogeneous electron transfer at the electrode surface with the desired synthetic transformation in solution [17]. Electrosynthesis is not only used to provide efficient, cleaner alternatives to known reaction processes but can also lead to the development of new reaction pathways that would otherwise not be accessible using traditional methods.

There are many opportunities to develop new electrosynthetic reactions to further broaden the applicability of electrochemical techniques to the preparation of complex organic targets. For example, a potentially important and underdeveloped area is electrosynthetic transformations that produce chiral products [18,19] directly based, for example, on chiral electrode surfaces [20] or on other strategies to electrochemically drive the formation of chirality. Such processes would be of considerable synthetic utility, and initial reports indicate that advances in the area will be made. For example, Kuhn and coworkers recently developed chirally imprinted metal electrodes [21] through electrodeposition of platinum in the presence of (S)-phenylethanol. The stereochemical information imprinted onto the electrode was then efficiently transferred to prochiral acetophenone, which was cathodically reduced/hydrogenated to (S)-phenylethanol (Figure 1.1). This case provides convincing evidence for chiral induction to work on molecular imprinted metal electrode surfaces, for example

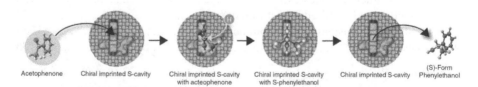

Acetophenone   Chiral imprinted S-cavity   Chiral imprinted S-cavity   Chiral imprinted S-cavity   Chiral imprinted S-cavity   (S)-Form Phenylethanol
with acteophenone   with S-phenylethanol

FIGURE 1.1 Schematic drawing of a chiral imprinted pocket in platinum binding acetophenone, which is then 2-electron 2-proton reduced to conform with the chiral pocket before releasing S-phenylethanol. (With permission from Wattanakit, C. et al., *Nature Commun.*, 8, 2087, 2017.)

during electrochemical hydrogenation. In future, molecular-level control over surface electrochemical reactions could offer tools to control chirality directly and to improve synthetic pathway engineering.

Further potential benefits from electrosynthesis could arise from the ability to operate in "continuous flow mode" rather than in "batch mode." Microreactor systems for electroorganic transformations with continuous flow have been developed which allow cascades of sequential reactions or also additional input of light energy in synthetic transformations [22]. Furthermore, some electroorganic reactions are energetically balanced and provide a potential benefit in terms of energy recovery. Processes can be either performed with power consumption or alternatively with simultaneous power generation as in "cogeneration" reactions [23]. Given the many factors associated with an electrochemical transformation and the many types of processes and reactors available, how is it possible to select the best conditions for a given organic synthetic challenge?

There are a number of general reviews on practical electroorganic chemistry [24] with information about the choice of apparatus, solvents, electrolytes, electrode materials, and reaction conditions. The choice of reaction conditions is very important in the success of the synthetic strategy and potentially crucial in making a process attractive to scale-up. Important considerations in the development of new electrosynthetic methods are simplicity in operating conditions, minimizing the need for specialist equipment, avoiding excess salts and additional reagents, avoiding complex purification protocol, minimizing process steps, and showing possibility for scale-up.

Recently, microfluidic reactor systems have been reported which provide new solutions that are highly beneficial and complementary to the requirements for continuous organic electrosynthesis [25]. In this regard, a recent review by Watts [26] details the use of microfluidic electrochemistry as a part of a wider range of microfluidic synthesis tools [27]. Yoshida's "cation pool" and "cation flow" technologies [28,29] offer examples for the versatile use of low-temperature formation of highly reactive intermediates (such as acyliminium cations, see Figure 1.2) coupled to

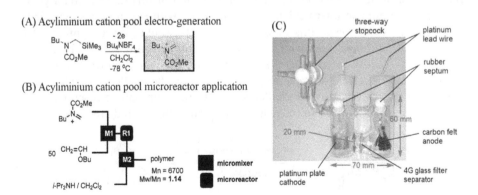

FIGURE 1.2 (A) Anodic electrolysis process for the formation of acyliminium cations. (B) Sequential reaction based on acyliminium-initiated polymer formation. (C) Depiction of a low-temperature H-type electrolysis "cation pool" cell. (With permission from Yoshida, J. et al., *Jpn. Chem. Rev.*, 2017, doi:10.1021/acs.chemrev.7b00475.)

microreactor technology for performing sequential reaction steps. These processes are continuous and can therefore be operated under optimized conditions and in parallel to increase scale. The formation of polymer products with defined molecular properties is only one example for a wider range of potential applications [30].

Microreactor methods have been developed further and, for example, multiphase flow [31] has been employed to allow synthesis and separation of products in continuous flow. Löwe and coworkers used microreactor techniques to produce ionic liquids continuously [32]. Commercial flow-through microreactor systems for organic electrosynthesis are now commercially available (see, e.g., [33]). These microreactors plug into a simple syringe pump system and use a simple battery or power supply to drive the continuous electrochemical reaction. An illustrative example of a flow-through microreactor process with a 200 µL volume undivided cell reactor using platinum and graphite electrodes was presented by Wirth and coworkers [34]. The reactor was developed to make benzothiazoles and thiazolopyridines by anodic cyclization of the corresponding N-aryl amides or thioamides, respectively (Scheme 1.1). Remarkably, this process gives excellent yields of the heterocyclic products in the absence of added electrolyte or any additional catalysts; ideal conditions for electroorganic synthesis. Such flow-through microreactors can be considered as "self-supported" when the reaction produces ionic charges in the confined space between the anode and the cathode.

1.1.1 Approaching Electroorganic Synthesis

One of the main potential advantages of electrosynthesis is that the reactivity of chemical reagents can be hugely expanded by oxidation or reduction steps that generate unusual reaction intermediates. Easy access to various types of intermediates from a single precursor allows new synthetic methods to be envisioned and exploited. This can be demonstrated by considering four (hypothetical) reaction pathways for the 4-nitrobenzyl radical (see Scheme 1.2) that could be generated through cathodic reduction of 4-nitrobenzylchloride. Further reduction of this radical would generate a reactive anionic intermediate that could react with electrophiles (e.g., CO_2). Alternatively, oxidation of this radical intermediate would give a highly reactive cation that would then react with nucleophiles. The radical may itself be used as the primary reactive intermediate, for example undergoing H-abstraction or radical dimerization. These examples highlight not only the challenge of planning a synthetic pathway but also the challenge of selecting appropriate reaction/reactor conditions that can provide selective access to each of the desired products.

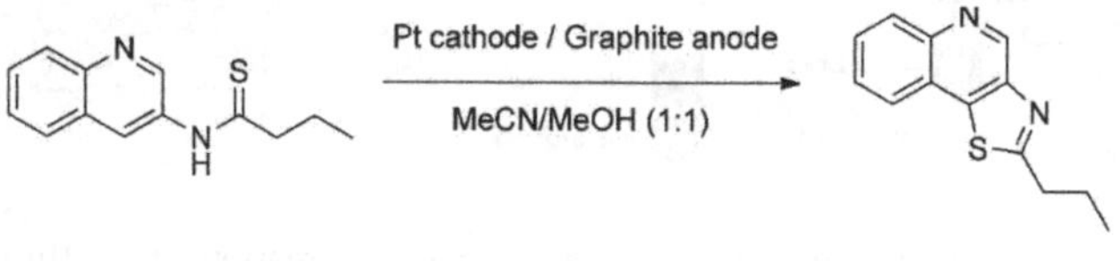

Pt cathode / Graphite anode
MeCN/MeOH (1:1)

SCHEME 1.1 Flow electrolysis example for an organic electrosynthesis reaction that requires no added salt or catalyst. (With permission from Folgueiras-Amador, A.A. et al., *Chem. Eur. J.*, 24, 487, 2018.)

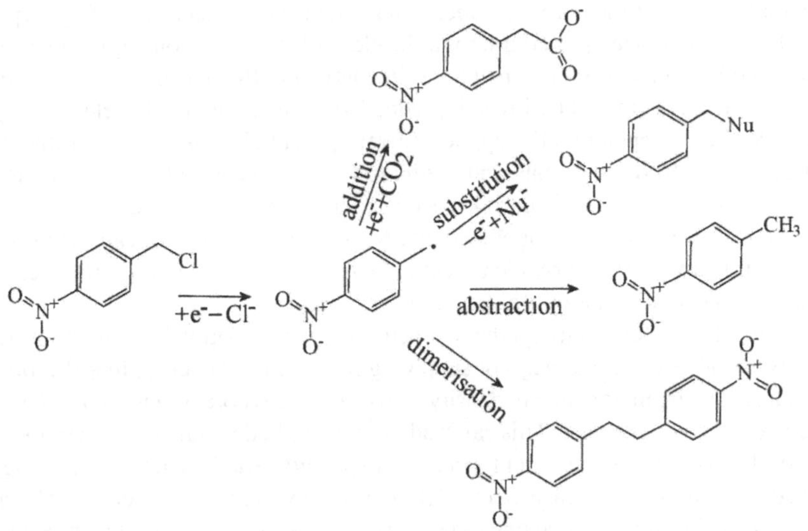

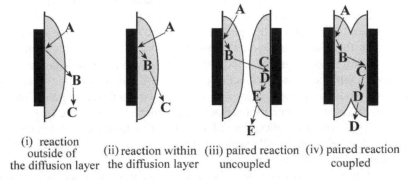

SCHEME 1.2 Reaction scheme (hypothetical) for the reduction of nitrobenzylchloride to lead into different reaction channels and products.

The hypothetical example reaction (Scheme 1.2) could potentially be performed in a number of practical ways. Figure 1.3 gives a schematic overview of the types of electrode processes occurring by (i) electron transfer and loss of the reactive intermediate for further reaction in the bulk solution, (ii) electron transfer with a reaction within the diffusion layer potentially followed by further electron transfer, (iii) anode and cathode both contributing to the overall paired process, and (iv) anode and cathode diffusion layers being coupled together to allow a sequence of reaction steps (anodic and cathodic) to occur locally within the diffusion layer and very rapidly. Most electrosyntheses would be performed as half-cell electrolysis (in

(i) reaction outside of the diffusion layer   (ii) reaction within the diffusion layer   (iii) paired reaction uncoupled   (iv) paired reaction coupled

FIGURE 1.3 Schematic drawing of electrode reaction sequences occurring outside/inside of the diffusion layer (the gray region) as well as with paired-uncoupled or paired-coupled diffusion layer conditions.

a divided cell, e.g., type i or ii), as microfluidic continuous electrolysis (e.g., type iii), or under self-supported paired microfluidic electrolysis conditions (type iv) as in the case of the Wirth reaction in Scheme 1.1. In contrast to the optimization of chemical yields in synthetic reactions involving added reagents, chemical yields in organic electrosyntheses depend on the applied current/potential, mass transport (the diffusion layer), electrode materials and configurations, solvent and electrolyte, as well as temperature. Overall, this introduces considerable added complexity and difficult choices. However, there is a lot of evidence for simple electrochemical methods to be successful in a variety of organic processes, even if all the possible parameters have not undergone significant optimization in every case.

Methodologies for electrosynthetic reactions can be broadly divided into different basic types of reactors and ways of supplying the electrical energy. In *galvanostatic* reactions, the current (or current density) is fixed and the potential is adjusted to drive the current through the cell. This can lead to high applied voltages and therefore possible problems with side product formation. In *potentiostatic* reactions, the potential is fixed (relative to a reference electrode) to the desired reaction potential, which can help perform highly selective reactions at the functional group of interest. However, as a consequence of the substrate being consumed, the current will decrease, slowing the rate of reaction and making it difficult to drive the process to completion. A number of different reactor setups can also be considered (Figure 1.4), including the use of sacrificial electrodes, undivided cells, divided cells, or bipolar cells.

Figure 1.4A shows a schematic depiction for the case of an undivided cell with a sacrificial electrode. For example, the sacrificial electrode could be based on either magnesium or zinc metal (for the anode), which would provide a negative potential to the cathode to drive a reduction. The applied potential is generated internally, and there is no need for externally applied power. Instead of the sacrificial electrode, a sacrificial reagent could be supplied to define the potential and drive a reaction at the working electrode. Positioning of the electrodes in this case is not critical, but agitation will be desirable. In Figure 1.4B, a simple undivided cell is considered with externally applied power and a noninterfering/inert (sacrificial or not) counter electrode. The applied potential can be adjusted to meet the requirements of the desired reaction. This is the most common configuration for electrolysis (see, e.g., the scalable allylic C–H oxidation by Horn et al. [35]), and conditions can be chosen that allow both electrodes to be operated within the same electrolyte solution (usually keeping the two electrodes apart and/or exploiting biphasic conditions [36,37] or imposing diffusional limiting conditions on the counter electrode process to push the potential into the limit of solvent electrolysis; e.g., microelectrode arrays can be employed). Both the anode and cathode may also be employed in paired electrolysis [5], so each can generate useful reaction products. Bringing the two electrodes closer together can be beneficial by reducing resistive losses in solution (providing conditions for "self-supported" reactions) and can act to further "couple" anode and cathode diffusion zones and processes.

Figure 1.4C shows a divided electrolysis cell with a salt bridge separating the individual half-cells. This is a commonly employed design for small-scale electrosynthesis (see, e.g., the H-cell in Figure 1.2) to avoid contamination of the working electrode compartment with products from the counter electrode compartment. This type of

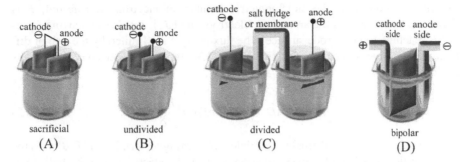

FIGURE 1.4 Schematic representation of types of electrolysis cells based on (A) undivided sacrificial, (B) undivided external power driven, (C) divided into two half-cells separated by membrane of salt bridge, and (D) bipolar with external driving power.

process therefore produces a lot of waste and, although avoiding some chemical complexity, is not particularly practical for laboratory electrosyntheses or scale-up. Stacks of electrodes can be switched in series in bipolar reactors (Figure 1.4D) that operate at higher voltage and with flow control. Bipolar stack reactors are designed with or without an additional membrane and are an important tool in industry-scale electrosynthesis [38].

For new electroorganic methods to be widely employed for synthesis, the electrochemical setup and equipment must be as widely accessible as possible. Ideally, an electrosynthetic method will (i) have a simple/cheap reaction setup, (ii) use no specialist glassware (e.g., use of undivided cells) or pumps, (iii) use a simple power source (e.g., battery or power source, avoiding expensive potentiostat equipment), (iv) use cheap and readily available electrodes (e.g., based on carbon or steel, etc.), (v) use solvent/electrolyte mixtures that are readily accessible and purifiable, (vi) have a convenient workup/separation of products from electrolyte, and (vii) generate minimum waste.

When designing a new electrosynthetic process, some important parameters are analogous to considerations in traditional organic synthesis (e.g., solvent, temperature, stoichiometry). However, there are also some important reaction parameters that are unique to electrosynthesis. For example, mass transport effects at the electrode surface are crucial in limiting the rate/yield of a particular process. If current flux exceeds the mass flux of molecules toward the electrode surface, then the reaction pathways can change, side products may form, and the synthesis may result in a mixture of products or proceed with a lower current yield.

1.1.2 MASS TRANSPORT

The current, I, flowing through an electrode during an electrochemical process is directly proportional to the flux of starting material reaching the electrode surface, as described for a reduction by Equation 1.1. Therefore, the current (or yield of product) may be higher for systems with a faster rate of mass transport (agitation).

$$I = -nFAJ_B \tag{1.1}$$

In this equation, n is the stoichiometric number of electrons transferred, F is Faraday's constant, A is the area of the electrode, and J_B is the flux of starting material B reaching the electrode surface. The flux is typically limited by the rate of diffusion in solution and can be described by Fick's first law of diffusion. This results in Equation 1.2.

$$I = -nFAD\frac{\partial[B]}{\partial x} = -nFAD\frac{[B]_{bulk}}{\delta} \tag{1.2}$$

In this equation, D is the diffusion coefficient of starting material B and $\partial[B]/\partial x$ is the concentration gradient of starting material B from the electrode into the bulk solution. This can be converted (approximately) into finite values for bulk concentration $[B]_{bulk}$ divided by the diffusion layer thickness δ as employed in the Nernst diffusion layer concept [39]. The diffusion layer thickness δ is a measure for the flux of molecules toward the electrode surface. The rate of a chemical reaction (occurring within or outside of the diffusion layer after electron transfer) can be expressed as reaction layer $\delta_{reaction}$ (here for a first-order rate constant k, Equation 1.3).

$$\delta_{reaction} = \sqrt{\frac{D}{k}} \tag{1.3}$$

The condition $\delta_{reaction} < \delta$ applies for fast reactions (occurring within the diffusion layer, see Figure 1.3), and the condition $\delta_{reaction} > \delta$ suggests a slow reaction (occurring outside of the diffusion layer, see Figure 1.3) with implication for the pathway of reactions that produce further electroactive intermediates. In highly viscous solvents, diffusion coefficients can be several orders of magnitude lower than for aqueous solutions, and consequently currents can be correspondingly small. One way to overcome the drawback of slow diffusion and low rates of conversion is to increase the rate of mass transport to the electrode surface using hydrodynamic techniques.

Hydrodynamic electrochemical methods encompass a range of techniques where forced convection is used to control the mass transport of material to the electrode surface. Forced convection is usually achieved either by moving the working electrode, for example in rotating disk and rotating ring-disk voltammetry, or by controlling the flow of solution across a static electrode, for example in wall-jet [40] or wall-tube [41] experiments, or in channel flow/microreactor electrochemistry [42]. The channel flow experiment is often carried out in rectangular flow ducts very similar to conditions found in many microfluidic reactors. Theory derived for channel flow is therefore often appropriate for processes in microreactor conditions. Simpler ways of agitation are also appropriate, e.g., agitation with a stirrer, to give an average or effective thickness of the diffusion layer $\delta_{effective}$. A "rocking disk" alternative to the rotating disk has been proposed, enabling hydrodynamic measurements to be recorded at a conventional inlaid-disk working electrode [43]. Nonuniformity of the diffusion layer may occur under flow conditions or at electrode edges [44]. An expression for the current (or the Nernstian diffusion layer thickness) under flow conditions can be derived to account for the type of flow and uniformity at an electrode of a particular geometry [45].

For highly viscous solvents (e.g., ionic liquids or polymer solvents), an alternative agitation technique has been proposed based on Couette flow in a microgap [46]. An electrode is positioned near to a rotating drum in the solution. The agitated solution is controllably passed across the surface of the electrode to allow the electrochemical reaction to proceed at very high flow rate controlled by rotation speed and microgap size (Figure 1.5). Electrolytic experiments were performed with highly viscous polyethylene-glycol (PEG) solvents.

1.1.2.1 Mass Transport and Computational Mechanistic Tools

With the diffusion layer thickness known (or controlled as the key process parameter), the reaction pathway in organic electrosyntheses can be modeled and understood in terms of a sequence of reaction steps and reaction intermediates (present either inside or outside the diffusion layer). For a known reaction mechanism, the yields and production of products can be quantified as a function of process parameters (e.g., stirring). For unknown reaction mechanisms, the current data as a function of diffusion layer thickness or as a function of added reagent can be modeled and fitted to experimental data. Hypothetical reaction mechanisms can be compared with real data sets. Commercial software packages (see, e.g., [47]) allow data analysis and prediction of reaction processes. As an example, in Figure 1.5, the one-electron oxidation of tetrachloro-N-hydroxyl-phthalimide (Cl_4NHPI) to tetrachloro-phthalimido-N-oxyl (Cl_4PINO) is shown to illustrate a homogeneous catalytic process. The Cl_4PINO radical is used as redox mediator and has the ability to abstract hydrogen atoms from C–H bonds [48], for example reacting with alcohols to give primarily aldehydes or ketones [49]. Comparison of current data (as a function of alcohol concentration) with simulation data for a mechanistic hypothesis revealed new insights into this reaction. Based on additional density functional theory (DFT) computational analysis, it was confirmed that C–H bond breaking is the rate-determining step, which was supported experimentally by the observation of a large primary kinetic isotope effect. The rate constant data for a range of alcohols were determined and correlated to DFT analysis of activation energy data (see Figure 1.6) to provide a predictive tool for other alcohols. Powerful DFT tools are now more widely available [50] to enable

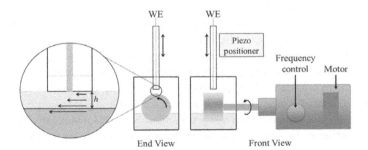

FIGURE 1.5 A schematic diagram of an apparatus suitable for performing hydrodynamic microgap voltammetry. The distance (h) between the working electrode (WE) and the drum may be controllably varied using a piezoelectric positioner. The frequency of rotation can also be varied. (With permission from Hotchen, C.E. et al., *ChemPhysChem.*,16, 2789, 2015.)

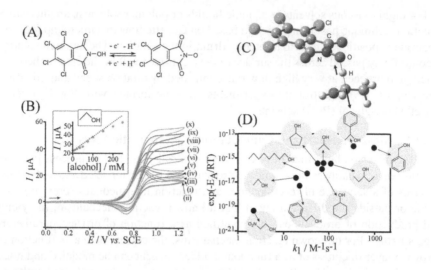

FIGURE 1.6 (A) One-electron oxidation of tetrachloro-*N*-hydroxyl-phthalimide (Cl₄NHPI) to tetrachloro-phthalimido-*N*-oxyl (Cl₄PINO). (B) Voltammetric data for the Cl₄PINO-catalyzed oxidation of ethanol as a function of concentration. (C) DFT transition state for C–H abstraction and (D) plot of experimental versus theoretical rate data. (With permission from Buckingham, M.A. et al., *Electroanalysis*, 2018, doi:org/10.1002/elan.201800147.)

further exploration of reaction pathways and reaction intermediates that might occur during electrode reactions.

1.1.3 Solvents and Electrolytes

A general rule (acknowledging that rules can be broken, e.g., see Scheme 1.1) in electrosynthesis is that the solution must provide a conductive pathway for current to flow and complete the electrical circuit between two electrodes. Typically, an ionic salt (also known as background electrolyte), such as KNO₃, is dissolved in a polar solvent, such as water, to provide the ionic conductivity required in the solution. The background electrolyte and the solvent should be both electrochemically inert and able to withstand high positive and negative potentials before being oxidized or reduced, respectively (the "solvent window"). Aqueous solutions offer a "green" approach [51]. However, water is not always a good solvent in organic electrosyntheses. Many organic starting materials are poorly soluble (or require surfactants to dissolve [52]) in aqueous solutions, making water a poor choice of solvent in these cases. The potential range of water is also narrow (due to hydrogen evolution and oxygen evolution) compared with many organic solvents. Many electroorganic transformations generate highly reactive intermediates that may react with water.

1.1.3.1 Organic Solvents

To overcome these problems, polar aprotic solvents, such as dichloromethane (CH₂Cl₂), acetonitrile (CH₃CN), or dimethylsulfoxide (DMSO), are commonly used as the solvents of choice for organic electrosynthesis due to their larger potential window [53]. The presence of an organic-soluble background electrolyte, such as tetrabutylammonium hexafluorophosphate (NBu_4PF_6), is required to avoid high resistance. In some cases, microgap electrochemical flow cells minimize losses due to resistance and therefore allow electrosyntheses even without intentionally added electrolyte [54]. The volatile organic solvents often exhibit low boiling points and evaporate quickly at room temperature, which can be hazardous but also beneficial in the workup and recovery of the solvent. In certain cases, the use of nonvolatile solvents, such as low-molecular-weight polymers (e.g., PEGs) and room-temperature ionic liquids (RTILs), can help address some of the challenges facing organic electrosynthesis.

1.1.3.2   Ionic Liquids

RTILs provide a modern class of solvents with interesting, unusual, and potentially useful properties in many areas of chemistry. They are defined as "materials composed of cations and anions, that melt around 100°C or below as an arbitrary temperature limit" [55]. Chum et al. reported "the first use of a room temperature, high Lewis acid molten salt system," which was based on the same 2:1 aluminum chloride: ethylpyridinium bromide mixture [56]. They performed voltammetric measurements without the addition of further electrolyte in the room temperature melt with a solvent potential window between −0.2 and +1.8 V (vs. Al reference electrode). However, the haloaluminate ionic liquids were highly water and oxygen sensitive, and it was necessary to handle the solutions in a glove box to avoid hydrolysis of the aluminum halide. The development of ionic liquids that do not contain aluminum halide salts improved the stability of the melts with respect to water and oxygen and caused a boom of research on this "new" class of non-haloaluminate or "room-temperature ionic liquids." The first of these air- and water-stable ionic liquids were based on the 1-ethyl-3-methylimidazolium ([EMIm]) cation [57]. Today, research continues to flourish in the field of RTILs with many areas of application including organic electrosynthesis [58]. Non-haloaluminate ionic liquids are typically composed of a bulky organic cation and an inorganic anion. Some of the more common RTILs are based on bulky N,N'-alkylimidazolium or N-alkylpyridinium cations, and a systematic nomenclature has been developed for these systems [59] (see Table 1.1). These cations can be used in combination with a range of readily available non-coordinating counter-anions.

In addition to the many combinations of cations and anions, RTILs have many advantageous chemical and physical properties that set them apart from typical organic solvents. RTILs are commonly reported to have low volatility, low combustibility, excellent thermal stability, intrinsic conductivity, good solvation properties for both polar and nonpolar compounds, high resistivity to oxidation and reduction, and the possibility that the solvent can be recycled. However, it should be emphasized that these properties are not necessarily a general trait for all RTILs. RTILs encompass an ever-increasing subset of chemical compounds, where each combination of cations and anions has unique properties. Consequently, there is frequently a large degree of variation observed in the properties of different RTILs [60]. It is therefore important to select an appropriate RTIL with desirable properties for the process to be studied or performed. The ionic conductivity of the ionic liquid is usually significantly lower compared with aqueous electrolytes and is dependent on ion size, anionic charge delocalization, viscosity, and density [61–64]. In general, cations

TABLE 1.1
Common Cations and Anions Used for RTILs and Their Commonly Abbreviated Names

Imidazolium	R_1	R_2	R_3	Abbreviation
	H	Methyl	H	[HMIm]
	Methyl	Methyl	H	[MMIm]
	Ethyl	Methyl	H	[EMIm]
	Propyl	Methyl	H	[PMIm]
	Hexyl	Methyl	H	[HexMIm]
	Butyl	Ethyl	H	[BEIm]
	Ethyl	Methyl	Methyl	[EMMIm]
	Butyl	Methyl	Methyl	[BMMIm]
	Hexyl	Methyl	Methyl	[HexMMIm]
	Methyl	Phenyl	Methyl	[MPhMIm]

N-Alkyl-N-methylpyrrolidinium	R_1	R_2	R_3	Abbreviation
	Ethyl	–	–	[Pyr$_2$]
	Propyl	–	–	[Pyr$_3$]
	Butyl	–	–	[Pyr$_4$]
	Pentyl	–	–	[Pyr$_5$]
	Hexyl	–	–	[Pyr$_6$]

N-Alkylpyridinium	R_1	R_2	R_3	Abbreviation
	Ethyl	–	–	[EtP]
	Butyl	–	–	[BuP]
	Pentyl	–	–	[PenP]
	Hexyl	–	–	[HexP]

Tetraalkylammonium	R_1	R_2	R_3	Abbreviation
	Methyl	Butyl	–	[Me$_3$BuN]
	Ethyl	Butyl	–	[Et$_3$BuN]
	Butyl	Methyl	–	[MeBu$_3$]
	Ethyl	Hexyl	–	[Et$_3$HexN]
	Ethyl	Octyl	–	[Et$_3$OctN]

Anions	Name	Abbreviation
	Hexafluorophosphate	[PF$_6$]$^-$

(*Continued*)

TABLE 1.1 (*Continued*)
Common Cations and Anions Used for RTILs and Their Commonly Abbreviated Names

Anions	Name	Abbreviation
(structure of BF_4)	Tetrafluoroborate	$[BF_4]^-$
(structure of triflate)	Triflate	$[OTf]^-$
(structure of dicyanamide)	Dicyanamide	$[N(CN)_2]^-$
(structure of triflimide)	Bis(trifluoromethylsulfonyl) amide (also known as triflimide)	$[NTf_2]^-$
(structure of BETI)	Bis(perfluoroethylsulfonyl) imide	$[BETI]^-$

Source: With permission from Hapiot, P. and Lagrost, C., *Chem. Rev.*, 108, 2238, 2008.

based on the imidazolium structure tend to exhibit the highest conductivities (~10 mS cm^{-1}), whereas quaternary ammonium salts are less conductive (~2 mS cm^{-1}) [59]. The conductivity of the ionic liquids can be increased by adding molecular solvents, such as water, and in some cases, Li$^+$ ions. The wider class of deep eutectic solvents [65] offers furthermore "green" alternatives to pure ionic liquids.

Highly pure and dry RTILs can exhibit a broad usable electrochemical window. For example, [BMIm][BF$_4$] has been reported to have an electrochemical potential window of up to 7 V, where the negative and positive limits were determined by the reduction of the cation and the oxidation of the anion, respectively [66]. However, the presence of impurities, in particular water, can significantly reduce the usable electrochemical potential window [67]. Ionic liquids are often hygroscopic, and the addition of even 3 wt.% water can severely decrease the usable electrochemical solvent window, in some instances by more than 2 V [68]. The viscosity of RTILs can be up to three orders of magnitude higher than for conventional molecular solvents [69]. Consequently, the rate of mass transport within ionic liquids can be slow, which could impact the turnover in electrosynthetic applications. However, hydrodynamic techniques and microreactor flow methods can be employed to circumvent the challenges posed due to slow mass transport. Particularly beneficial are ionic liquids in "green" anodic fluorination electrosynthesis [70]. Another important consideration for ionic liquid application is based on gas solubility [71] leading to enhanced electroorganic reactivity. The electrosynthetic reduction of carbon dioxide offers promise and provides an example of ionic liquids beneficially enhancing processes [72,73]. For example, at silver electrodes [74], the reduction of carbon dioxide in various types of ionic liquids leads to both (A) specific catalytic effects due to the imidazolium radical and the imidazolium-carbene and (B) general effects due to bulky cation effects on the double-layer structure where surface transition states can be energetically favored (Figure 1.7).

1.1.3.3 Polymer Solvents

PEG is generally considered to be an inexpensive, environmentally benign, nontoxic, nonvolatile, highly viscous polymer, which has many interesting properties including biofouling resistance when attached to surfaces [75–77]. PEG is also known to absorb carbon dioxide (CO_2) well, which could suggest potential applications as a solvent for the electrochemical capture and conversion of carbon dioxide [78]. The capture solvent monoethanolamine has been successfully used for carbon dioxide reduction [79], and carbon dioxide reduction in polymer gels has been reported [80]. Although the use of PEG as a solvent in electrosynthetic processes is in its infancy, the solvent could see many future applications in this field.

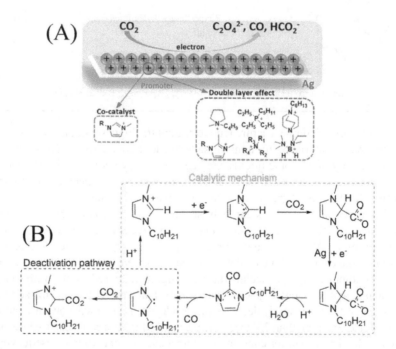

FIGURE 1.7 (A) Schematic drawing and (B) reaction scheme for carbon dioxide reduction in ionic liquid and at a silver electrode in the presence of a cation layer acting as the promotor. In the case of imidazolium cations, additional co-catalysis effects are possible. (With permission from Zhao, S.F. et al., *J. Phys. Chem. C*, 120, 23989, 2016.)

1.2 METHODOLOGY FOR DIRECT TRANSFORMATIONS

1.2.1 ANODIC PROCESSES

The Kolbe carboxylate oxidation is a historical [81] but also very useful reaction, which is widely explored as a convenient tool to generate anodic radical intermediates at platinum electrode surfaces [82]. Mild generation of radical intermediates leads to a range of possible coupling and cross-coupling products. For example, Tajima et al. [83] have employed the Kolbe reaction to form C–C dimers from carboxylic acids. They use a simple undivided cell with 10 mL of solvent (methanol–acetonitrile) without electrolyte but with beads of silica-immobilized piperidine base, two platinum electrodes, and constant current during electrolysis (see Figure 1.8). This process relies on the solid-phase base deprotonating small amounts of methanol to provide sufficient conductivity and to provide the basic conditions needed for the Kolbe coupling. As a result, the products do not require separation from electrolyte or salt,

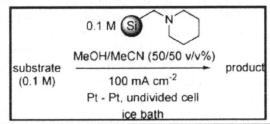

entry	substrate	electricity (faraday mol^{-1})	product	yield (%)
1	$C_7H_{15}CO_2H$	3	$C_{14}H_{30}$	90
2	$C_9H_{19}CO_2H$	3	$C_{18}H_{38}$	99
3	$MeO_2C(CH_2)_2CO_2H$	4	$MeO_2C(CH_2)_4CO_2Me$	91
4	$MeO_2C(CH_2)_4CO_2H$	4.5	$MeO_2C(CH_2)_8CO_2Me$	88
5	$MeO_2C\diagdown\diagup CO_2H$	2.5	$MeO_2C\diagdown\diagup CO_2Me$	91
6	(benzyl CO_2H)	1.5	(bibenzyl)	44
7	(pentafluorobenzyl CO_2H)	2	(perfluorinated bibenzyl)	52

FIGURE 1.8 Reaction scheme and yields for the Kolbe oxidation coupling reactions that are carried out in an undivided cell and without intentionally added salt or electrolyte. Instead, silica beads with surface-immobilized base are employed to generate sufficient conductivity. (With permission from Kurihara, H. et al., *J. Org. Chem.*, 73, 6888, 2008.)

and the process is much more convenient. The solid-phase base reagent is simply removed by filtration and can be recycled.

Kolbe-related processes have been recently reported for 2-arylbenzoic acid derivatives where the anodically generated carboxylate radical intermediate undergoes intramolecular cyclization onto the adjacent aryl ring to form substituted chromenones in high yield [84]. Zhang and coworkers have reported similar methodology for the dehydrative lactonization of carboxylic acids onto internal C–H bonds to form a wide range of substituted lactones (Figure 1.9) [85]. The process uses a simple undivided cell and either platinum or carbon electrodes under constant current in acetonitrile/methanol with NBu_4BF_4 electrolyte. Importantly, this simple procedure has been demonstrated to be highly scalable using inexpensive carbon electrodes to form 33 g of chromenone (see Figure 1.9) without the need for chromatographic purification.

For these types of reactions, there is no need for a complicated power source due to constant current/galvanostatic conditions. Further examples of electroorganic reactions employing a battery power source have been suggested by Moeller and coworkers for intramolecular anodic alkene coupling reactions [86].

This work was expanded into electroorganic synthesis driven directly by simple solar cell power sources [87]. A related development is nanoparticulate "solar power sources" shown, for example, to convert aliphatic alcohols into esters [88]. Recent reviews of this emerging field have appeared [89,90], highlighting the opportunities from the design of molecular [91] and of semiconductor nanoparticulate photocatalytic systems. For example, a platinum-loaded tungsten oxide photocatalyst (or with lower-efficiency platinum-loaded titanium oxide) can be used for the selective oxidation of benzene into phenol (Figure 1.10), with water acting as the primary source of reactive hydroxyl radicals [92].

Wright and coworkers have developed a range of anodic intramolecular annulations of silyl enol ethers with furans using undivided cells under constant current and a carbon anode [93–96], which has been further applied to the synthesis of complex tricyclic products [97]. An anodic sp^3 C–H amination of various saturated heterocycles using azoles has been reported by Lei and coworkers [98]. The reaction uses platinum electrodes and $n\text{-}Bu_4NBF_4$ as electrolyte in acetonitrile at 80°C in an undivided cell. The reaction works for a range of substrates to form the amination products in generally high yields and can be performed on a preparative scale. For example, the reaction of benzotriazole with tetrahydrofuran (THF) was performed on a 10 mmol scale to form 1.2 g of the product (Scheme 1.3). Mechanistic studies suggest anodic generation of the triazole radical followed by oxidative coupling to THF.

Ley and coworkers have used a commercial continuous flow electrosynthesis microreactor system for the electrochemical alkylation of acetonitrile using benzylic substrates [99]. The reactor system uses a platinum anode and a stainless steel cathode and $n\text{-}Bu_4NPF_6$ as electrolyte. The use of methanesulfonic acid as an additive was essential for maintaining homogeneity and process stability to allow for continuous production of the amide products (Scheme 1.4).

Existing synthetic transformations based on chemical reagents can often be "reengineered" and the chemical reagent replaced with a suitable electrode process.

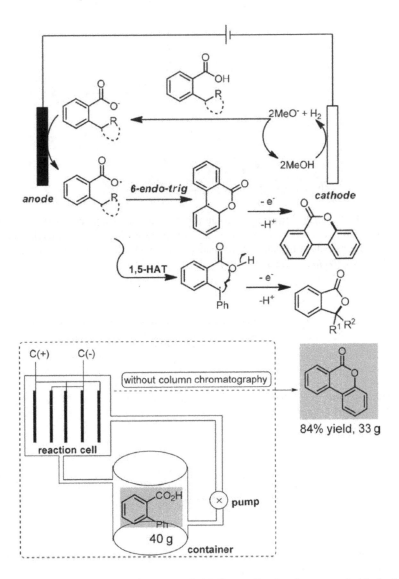

FIGURE 1.9 Reaction scheme for an anodic Kolbe cyclization (accompanied by hydrogen evolution in an undivided cell) for sp² and sp³ cyclization. A reactor for multi-gram electrosynthesis is suggested. (With permission from Zhang, S. et al., *Org. Lett.*, 20, 252, 2018.)

For example, the radical C–H trifluoromethylation of nitrogen heterocycles using zinc sulfinates as radical precursors can be performed using super-stoichiometric *tert*-butyl-hydroperoxide (TBHP) as an initiator. However, Baran and coworkers showed that the use of a simple electrochemical oxidation leads to a cleaner and more efficient process, with generally higher yields of the synthetically valuable trifluoromethylated *N*-heterocycles [100]. The reaction is performed under constant current conditions in a divided cell using a carbon cloth anode in a mixture of Et₄NClO₄ and DMSO (Figure 1.11).

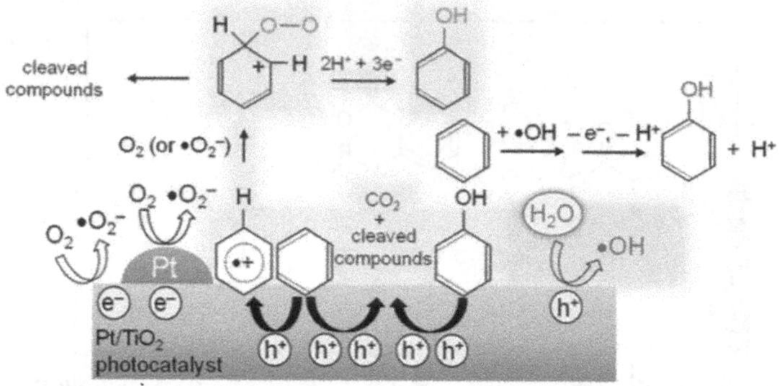

FIGURE 1.10 Schematic drawing of an illuminated TiO_2 semiconductor particle with platinum deposit. Holes are driven to the semiconductor surface where they hydroxylate benzene, whereas electrons collect in platinum to reduce oxygen. (With permission from Tomita, O. et al., *Catal. Sci. Technol.*, 4, 3850, 2014.)

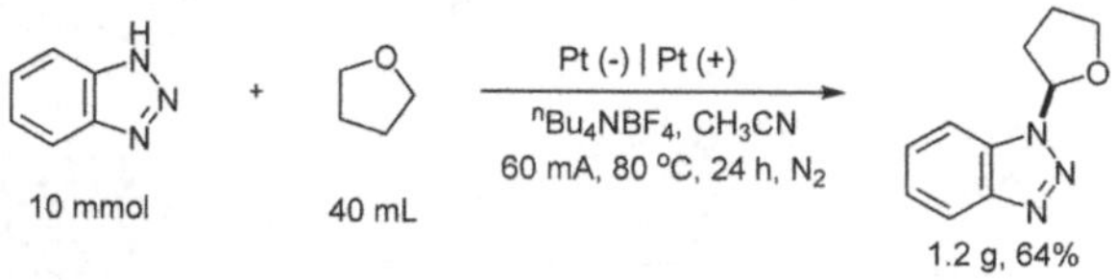

SCHEME 1.3 Electrosynthetic formation of an aminated THF. (With permission from Wu, J.W. et al., *ACS Catal.*, 7, 8320, 2017.)

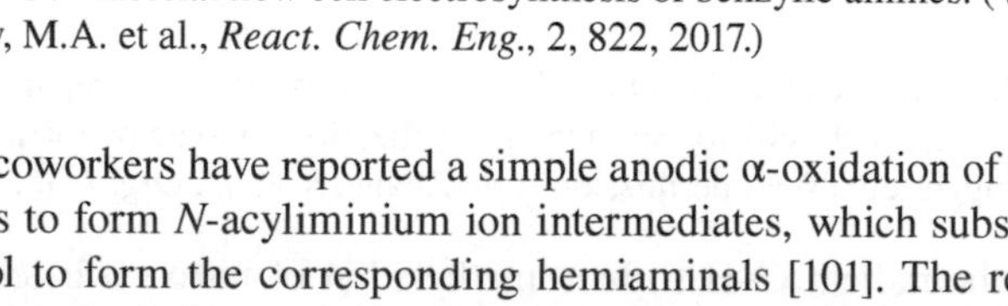

SCHEME 1.4 Commercial flow cell electrosynthesis of benzylic amines. (With permission from Kabeshov, M.A. et al., *React. Chem. Eng.*, 2, 822, 2017.)

Aubé and coworkers have reported a simple anodic α-oxidation of complex polycyclic lactams to form *N*-acyliminium ion intermediates, which subsequently react with methanol to form the corresponding hemiaminals [101]. The reaction is performed with standard glassware (undivided), with a simple battery power source, and works well with pencil graphite as the electrodes. A range of substrates were tested, forming the products in generally good yield using either methanol/$LiClO_4$ or methanol/NBu_4OTs as the solvent and electrolyte (Scheme 1.5).

Xu and coworkers reported an intramolecular oxidative amidation of alkenes using a simple undivided cell and a carbon anode and Pt cathode, forming a range of substituted lactams in good yields [102]. They also managed to use anodic

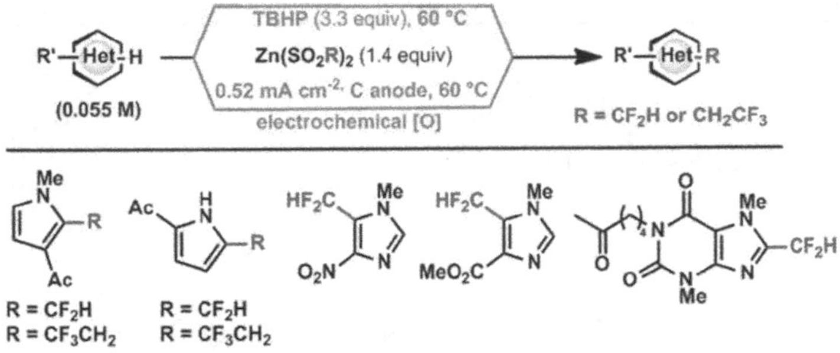

FIGURE 1.11 Heteroaromatic C–H alkylation based on sulfonate oxidation to replace the traditional *tert*-butyl-hydroperoxide (TBHP) process. (With permission from O'Brien, A.G. et al., *Angew. Chem. Int. Ed.*, 53, 11868, 2014.)

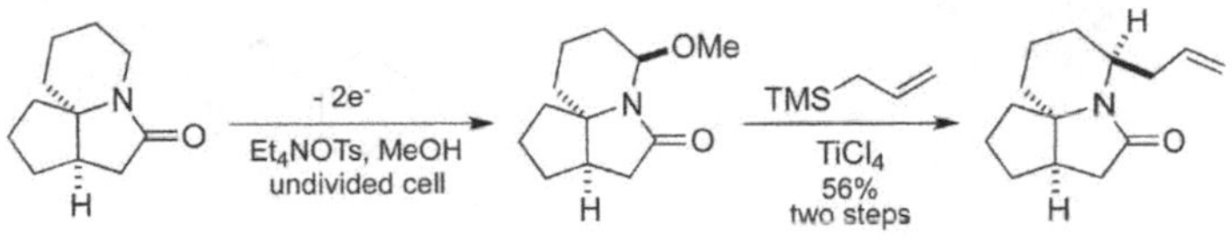

SCHEME 1.5 Electroorganic methoxylation of a C–H bond to introduce functionality. (With permission from Frankowski, K.J. et al., *Angew. Chem. Int. Ed.*, 54, 10555, 2015.)

oxidation to perform challenging aryl C–O bond formation through cyclization of *N*-benzylamides (see Figure 1.12 [103]). The process is performed in an undivided cell at constant current using a carbon anode with a platinum cathode and generates hydrogen as the only by-product. The proposed mechanism involves initial anodic oxidation of the electron-rich aryl ring, followed by regioselective intramolecular cyclization of the amide onto the para-position. Further oxidation followed by proton removal regenerates aromaticity and forms the product.

Anodic oxidation can also be used to initiate radical chain reactions with high efficiency. For example, Imada and coworkers [104,105] demonstrate that the Diels–Alder reaction between a styrene derivative and isoprene gave product in 80% yield on a multi-gram scale, with a current efficiency of 8000% (Figure 1.13). The process is carried out at constant potential using carbon felt electrodes in nitromethane and with $LiClO_4$ as electrolyte. The proposed mechanism involves initial anodic oxidation of electron-rich styrene into the corresponding radical cation, which undergoes a Diels–Alder cycloaddition with isoprene to form an intermediate radical cation. This species is sufficiently oxidizing to react with the starting material directly, forming the product and propagating the radical chain process. This reaction therefore represents a highly efficient electrosynthetic procedure in which only a catalytic quantity of electrons is required from the external power source.

The diversity of possible reaction types in anodic functional group replacement reactions is further illustrated with a recent work by Huang and coworkers [106], who reported a novel intermolecular decarboxylative sulfonylation of cinnamic acids with sulfonylhydrazides (Scheme 1.6). The reaction occurs in an undivided

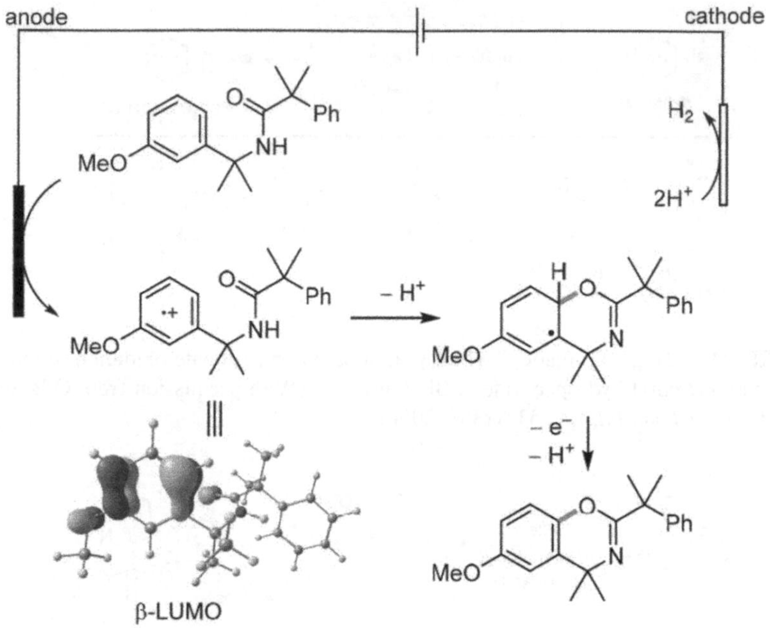

FIGURE 1.12 Reaction scheme for the intramolecular cyclization of amides to give C–O aromatic functionalization. LUMO, lowest unoccupied molecular orbital. (With permission from Xu, F. et al., *Org. Lett.*, 19, 6332, 2017.)

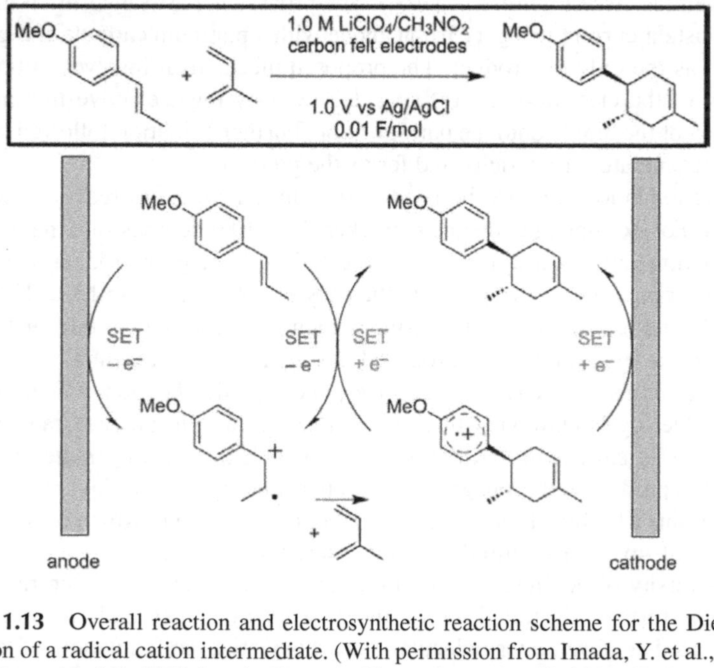

FIGURE 1.13 Overall reaction and electrosynthetic reaction scheme for the Diels–Alder cyclization of a radical cation intermediate. (With permission from Imada, Y. et al., *Beilstein J. Org. Chem.*, 14, 642, 2018; Imada, Y. et al., *Chem. Commun.*, 53, 3960, 2017.)

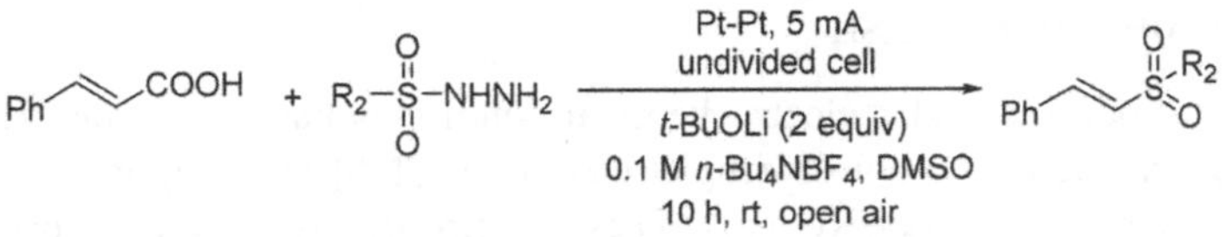

SCHEME 1.6 Undivided cell electrosynthesis of vinylsulfones from cinnamic acid and sulfonylhydrazides. (With permission from Zhao, Y. et al., *J. Org. Chem.*, 82, 9655, 2017.)

cell using platinum electrodes at constant current, using lithium *tert*-butoxide as base, n-Bu$_4$NBF$_4$ as electrolyte, and DMSO as solvent. The reaction scope has been demonstrated for a range of substituted cinnamic acids and sulfonylhydrazides bearing various nonparticipating functional groups, forming the substitution products in good yields. Mechanistic control experiments suggest that sequential two-electron anodic oxidation and deprotonation of the sulfonylhydrazide releases N$_2$ and generates a reactive sulfonyl radical that adds to the cinnamic acid. Facile oxidative decarboxylation then generates the observed vinyl sulfone products.

The oxidative dimerization of substituted indolizine nitrogen heterocycles was observed by Han and coworkers [107] in an undivided cell system with carbon anode and platinum cathode at constant current (Figure 1.14). The reaction produced good yields for a range of substrates under mild conditions. The proposed mechanism suggests anodic radical cation formation followed by dimerization, with subsequent cathodic reduction generating hydrogen as the only by-product.

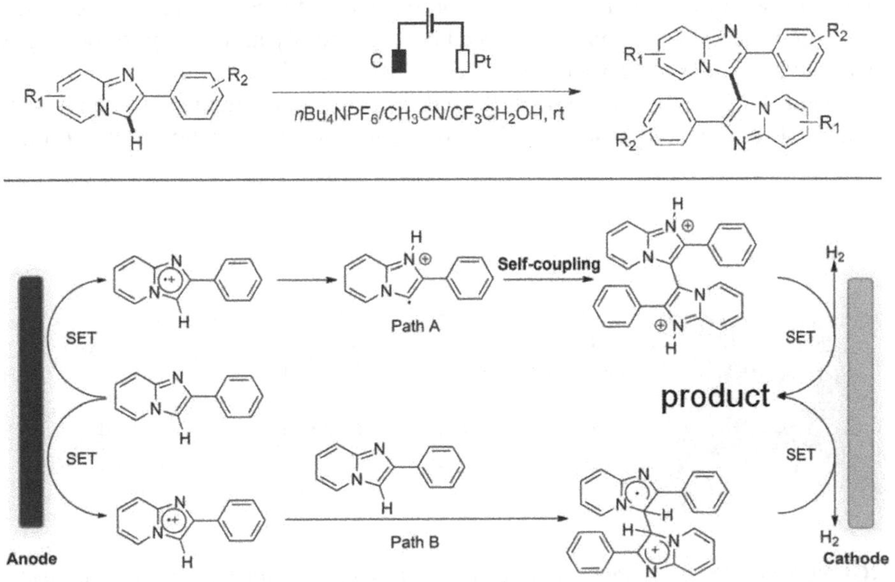

FIGURE 1.14 Dimerization of nitrogen heterocycles under undivided cell conditions. (With permission from Gao, Y.Y. et al., *Green Chem.*, 20, 583, 2018.)

1.2.2 CATHODIC PROCESSES

Reduction processes under electrochemical conditions have for many years been performed on mercury pool or droplet electrodes [108] as exceptionally versatile negative potential electrode systems. These electrodes are unable to form surface hydrides and therefore provide an extended negative potential range even under protic solvent conditions. However, mercury is not particularly convenient to handle in the laboratory, poses decontamination problems, and is probably now impossible to implement as cathode material in industry processes. Therefore, other types of electrode materials are needed. Many standard reduction processes including carbonyl reduction, nitro-group reduction, or dehalogenation are readily performed on carbon or graphite electrodes [109]; however, there is still a need for the development of a wider range of new cathode materials.

Waldvogel and coworkers have recently developed a new class of cathode materials based on leaded bronze [110]. This material proved crucial in the reductive dehalogenation of dibromo-cyclopropanes on the multi-gram scale in a batch reactor and on the industrial scale under flow conditions [111]. Figure 1.15 shows the 1-L batch reactor with (a) the leaded bronze cathode, (b) the porous ceramic separated anodic compartment, (c) the platinum anode, and (d) a high-power stirrer. The cathodic dehalogenation occurs in dimethylformamide (DMF) solvent with high yield. The amount of electricity supplied is doubled to minimize the formation of the mono-bromo intermediate.

A special class of reduction processes is based on Birch-type [112] reaction conditions (with solvated electrons), for example, in liquid ammonia [113] or in ethylenediamine [114]. This type of process (due to the formation of solvated electrons) requires a divided cell or sacrificial anodes based, for example, on magnesium [115]. The cathode can be metal-, carbon-, or boron-doped diamond. Under these conditions, aromatic hydrocarbons are reduced homogeneously in the 1,4 position to give thermodynamically unstable 1,4-hexadienes (Scheme 1.7).

A field of high research activity is the reduction of carbon dioxide (which is highly surface site and condition sensitive) and the reductive carboxylation of unsaturated

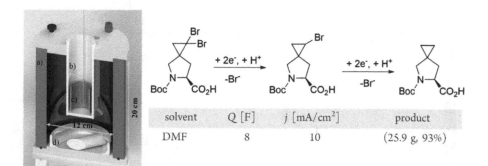

solvent	Q [F]	j [mA/cm^2]	product
DMF	8	10	(25.9 g, 93%)

FIGURE 1.15 Cross-sectional drawing of a 1-L batch reactor (divided) with (a) the leaded bronze cathode, (b) the porous ceramic separated anodic compartment, (c) the platinum anode, and (d) a high-power stirrer. The cathodic de-bromination in DMF affords cyclopropane derivatives. (With permission from Gütz, C. et al., *Org. Proc. Res. Dev.*, 19, 1428, 2015.)

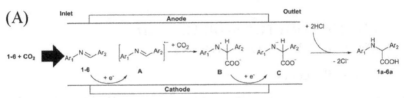

SCHEME 1.7 Birch reduction based on solvated electrons reducing aromatic systems in 1,4 position.

precursor molecules or halogenated compounds [116,117]. As an example, the electrocarboxylation of imines to amino acids in a microreactor system has been demonstrated by Atobe and coworkers (see Figure 1.16 [118]). The process has been optimized to work in THF solvent at a carbon cathode. The microfluidic reactor is based on rectangular flow driven by a syringe pump, and carbon dioxide is present as a reagent to capture the imine radical anion intermediate.

The combination of cathodic dehalogenation with in situ carboxylation using CO_2 was studied by Luo and coworkers [119] in DMF solution at silver electrodes. The reaction was driven with potential control versus a sacrificial magnesium electrode using $n\text{-Bu}_4NCl$ as electrolyte. While *ortho-*, *meta-*, and *para-*dichlorobenzene all

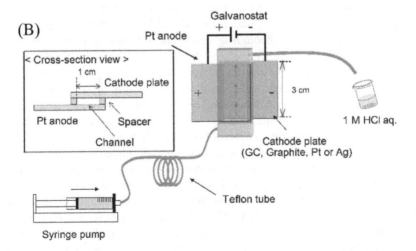

FIGURE 1.16 (A) Reaction scheme for a flow-through imine reduction followed by CO_2 coupling. (B) Microfluidic reactor system with platinum anode and different types of cathodes. (With permission from Qu, Y. et al., *React. Chem. Eng.*, 2, 871, 2017.)

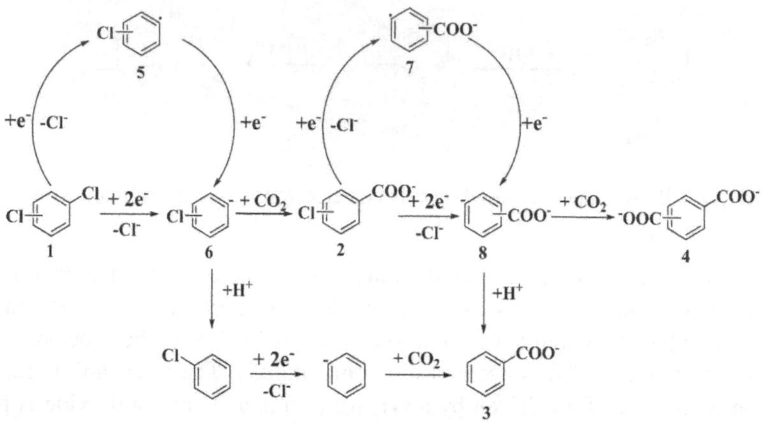

FIGURE 1.17 Reaction scheme for the cathodic dehalogenation and CO_2 addition for dichlorobenzenes. (With permission from Luo, P.P. et al., *Catalysis*, 7, 274, 2017.)

underwent monocarboxylation, the second dehalogenation/carboxylation step only worked for *para*-dichlorobenzene (Figure 1.17).

1.3 METHODOLOGY FOR INDIRECT AND CATALYTIC TRANSFORMATIONS

1.3.1 ANODIC PROCESSES

Often electrochemical mediators are useful in controlling/directing electrochemical reactions. Direct electroorganic synthesis relies on direct electron transfer and substrate interaction with the electrode surface, but indirect or catalytic processes occur homogeneously with a "shuttle" (such as Ce(IV) [120]) delivering the charges to the substrates of interest [121]. Although at first sight this may seem wasteful, redox mediators can be simple and effective. Another common strategy is immobilization of the redox mediator at the electrode surface [122]. A potential benefit of redox mediators is that the heterogeneous electron transfer step of direct electrosynthesis (usually slow and mass transport dependent) is replaced with a homogeneous electron transfer between the mediator and substrate (independent of the diffusion layer of the electrode). Reviews cover various classes of redox mediators (inorganic, organic, and biological [123,124]), and here only some illustrative examples are provided.

An advantage of using redox mediators is that bonds/functional groups with usually inaccessible redox potentials (e.g., outside of the working range of common solvents) may be activated. For example, Baran and coworkers have reported a C–H activation method for the oxidation of remote C–H bonds under readily accessible electrosynthetic conditions [125]. The reaction is performed in an open beaker or bucket (in air) using stoichiometric quinuclidine as an inexpensive redox mediator. The reaction occurs at constant current using Me_4NBF_4 as electrolyte and HFIP as solvent using a reticulated vitreous carbon anode (a porous form of glassy carbon) with a nickel foam cathode generating hydrogen as a side reaction. Figure 1.18 illustrates the need for this unusual solvent to avoid the highly active redox mediator attacking the solvent rather than the nonactivated hydrocarbon.

Figure 1.19 shows a reaction summary and the reaction conditions. The scope of this reaction is remarkable with primary, secondary, and tertiary C–H bonds of

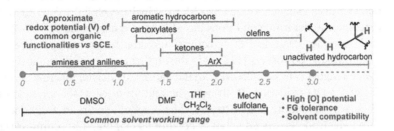

FIGURE 1.18 Scheme of solvent stability over the potential scale and requirements for the activation of unactivated hydrocarbon. (With permission from Kawamata, Y. et al., *J. Am. Chem. Soc.*, 139, 7448, 2017.)

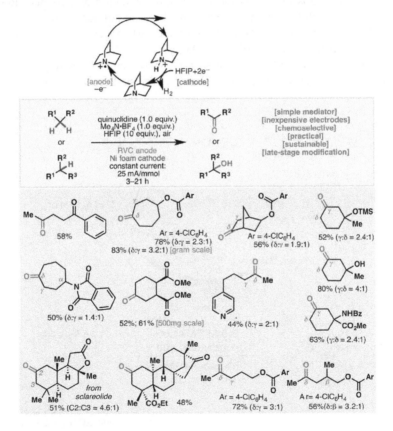

FIGURE 1.19 Reaction summary and reaction conditions for the quinuclidine-mediated oxidation of nonactivated C–H functional groups. RVC, reticulated vitreous carbon. (With permission from Kawamata, Y. et al., *J. Am. Chem. Soc.*, 139, 7448, 2017.)

complex molecules bearing a variety of nonparticipating functional groups undergoing highly regioselective oxidation into the corresponding alcohols or ketones. The synthetic utility and scalability of this process has been demonstrated through the selective C–H oxidation of the sesquiterpene natural product sclareolide on a 50 g scale. The efficient electrosynthetic oxidation of allylic C–H bonds into α,β-unsaturated ketones using catalytic tetrachloro N-hydroxyphthalimide (20 mol%) as a redox mediator under mild conditions has also been reported [126].

A Pd(II) complex-catalyzed C–H activation/oxidation process has been developed by Sanford and coworkers [127]. It is shown that a commonly employed hypervalent iodine reagent can be replaced with the anodically driven reaction proceeding via Pd(II) to Pd(IV) oxidation. Both benzylic and aromatic C–H activation by Pd followed by acetylation are demonstrated. The reaction occurs in a divided cell at graphite rod electrode in stirred solution conditions (Figure 1.20). Similar types of reactions for aliphatic C–H systems were reported by Mei and coworkers [128].

In contrast to the work with homogeneous redox mediator systems, Berben, Little, Francke, and Johnson reported carbon electrodes that are covalently modified with imidazole-based redox mediators for the versatile oxidation of alcohols and ethers [129]. A wide range of reactions are performed in an undivided cell in acetonitrile solvent. The redox mediator is "clicked" onto the carbon electrode surface and then employed to transfer electrons (Figure 1.21).

A simple hyper-iodine-based process with iodoaromatic electrolyte is another example of a homogeneous redox mediator reaction. Broese and Francke [130]

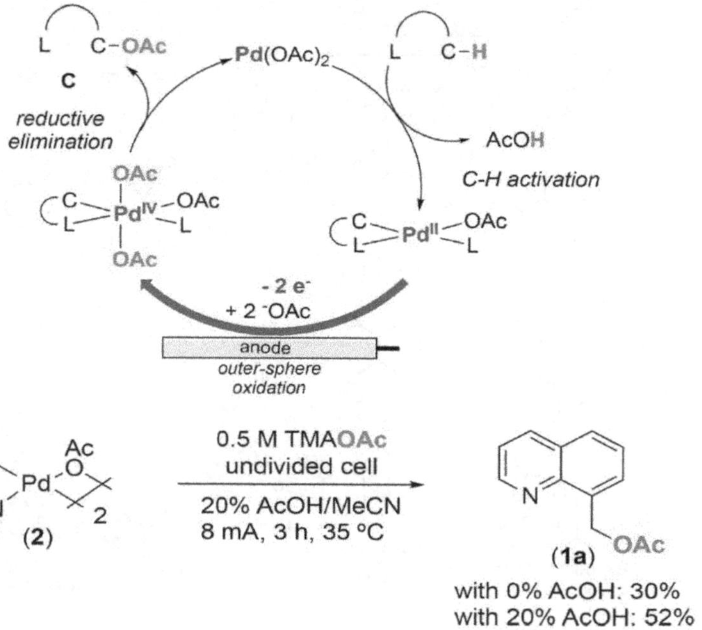

FIGURE 1.20 Reaction scheme for the benzylic or aromatic C–H activation with electrochemically driven Pd catalyst. (With permission from Shrestha, A. et al., *Org. Lett.*, 20, 204, 2018.)

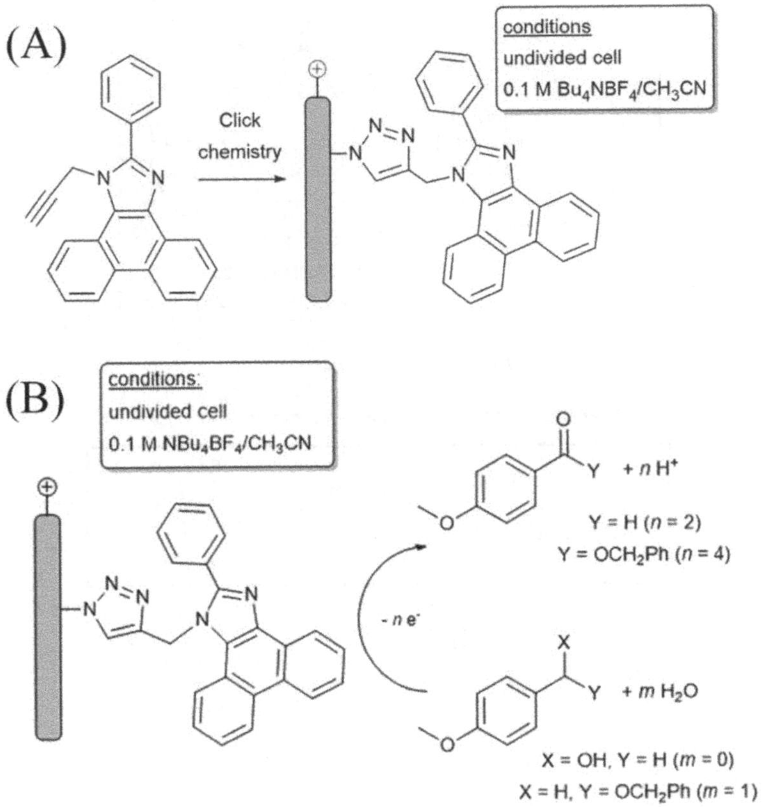

FIGURE 1.21 Reaction scheme for (A) immobilization and (B) application of a surface-bound redox mediator in oxidation of alcohols. (With permission from Johnson, B.M. et al., *Chem. Sci.*, 8, 6493, 2017.)

developed an inter- or intramolecular oxidative coupling process between amides and aryl rings (aromatic amidation) based on an in situ-generated catalytic iodine (III) system (Figure 1.22). The electrolyte is employed as mediator/catalyst, and the solvent hexafluoro-isopropanol (HFIP) is involved in the reaction (Figure 1.21). Recovery and reuse of the mediator/electrolyte are demonstrated.

Little, Zeng, and coworkers [131] suggested the ammonium iodide/iodine-mediated synthesis of a range of sulfonamides (Figure 1.23). The process appears attractive as it is performed in an undivided cell and using a carbon/graphite anode. The underlying mechanism is relatively complicated but based mainly on the sulfinate radical intermediate reacting with the amine-releasing iodide. Yields are good, and the scale-up of this process is reported.

The important lignin depolymerization process to give a bio-resourced feedstock of precursor molecules was reported by Stephenson [132]. *N*-Hydroxyphthalimide in the presence of lutidine base was the catalyst for hydrogen atom transfer, which allowed natural lignin to break up. A second reaction step based on photocatalytic lignin depolymerization was applied to the same solution to yield valuable products (Figure 1.24).

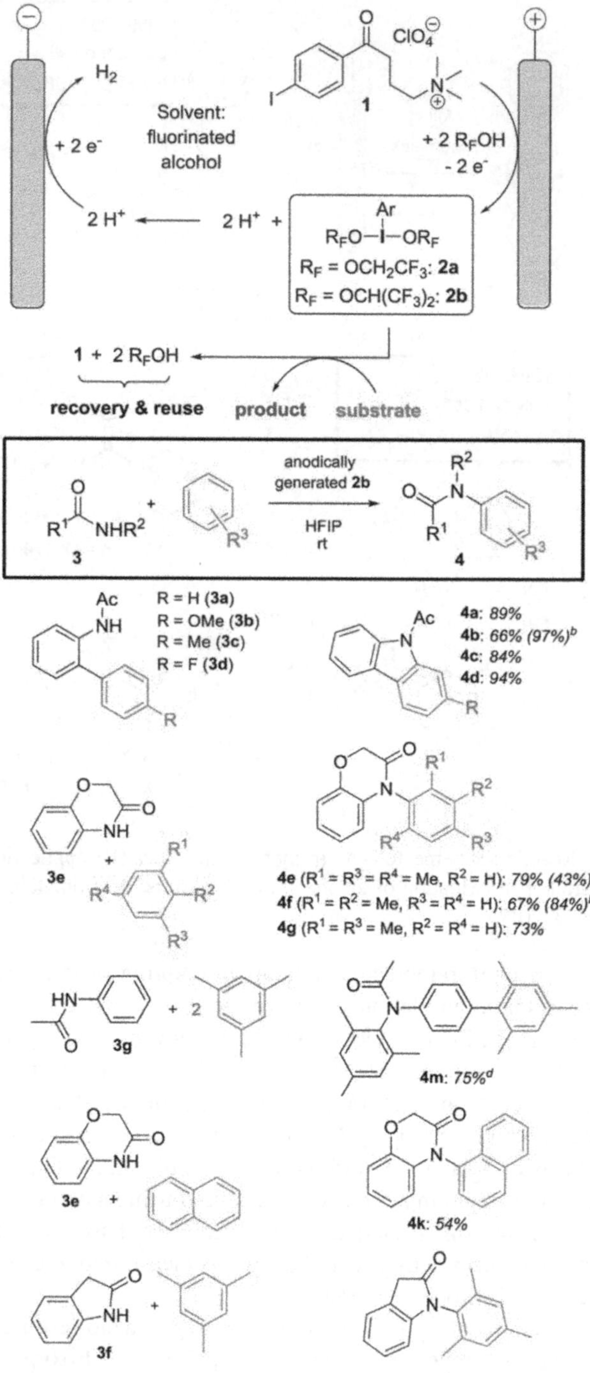

FIGURE 1.22 Reaction scheme for aromatic amidations driven by hyper-iodine redox mediators. (With permission from Broese, T. and Francke, R., *Org. Lett.*, 18, 5896, 2016.)

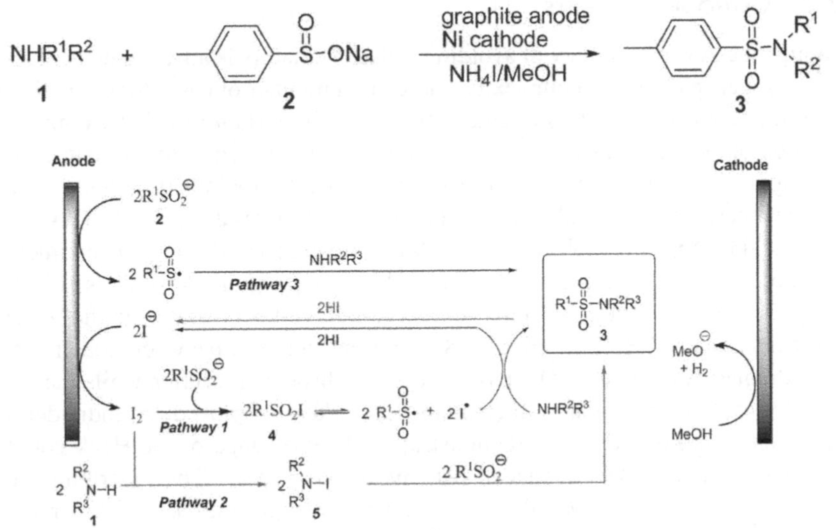

FIGURE 1.23 Schematic reaction for the formation of sulfonamides based on sulfonate radicals. (With permission from Jiang, Y.Y. et al., *J. Org. Chem.*, 81, 4713, 2016.)

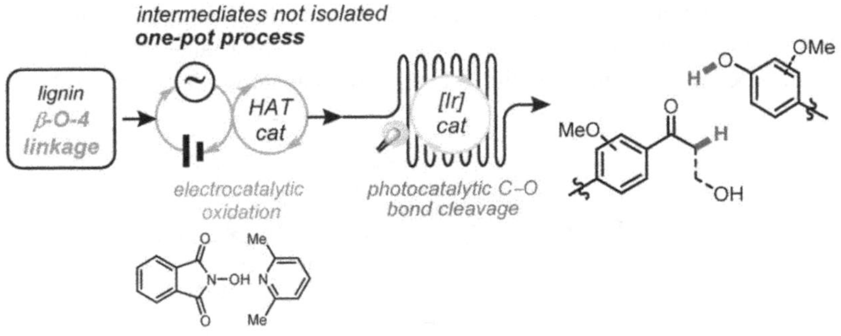

FIGURE 1.24 Schematic description of the two-step lignin depolymerization process developed by Stephenson and coworkers. (With permission from Bosque, I. et al., *ACS Central Sci.*, 3, 621, 2017.)

Ackermann developed a cobalt-catalyzed oxidation of aryl rings based on a simple divided cell process with constant applied current [133]. Lin and coworkers developed manganese-based metal complex redox catalysis that can be employed in the dichlorination of alkenes (in multi-gram scale [134]) and the trifluoromethylation of alkenes [135]. Ferrocene-based redox catalysis has been reported, for example, by Xu for intramolecular radical cyclization [136] and for difluoromethylarylation of alkynes [137]. Ferrocene was also used for the generation of benzylic radicals from boronate derivatives [138] and for catalyzing [3+2] cycloaddition processes [139] to form heterocyclic products in undivided cells under constant current conditions.

1.3.2 CATHODIC PROCESSES

Although the general strategy of avoiding redox mediators is advantageous in terms
of process parameters and complexity, there are a number of cases for cathodic processes where redox mediators or catalysts play an important role. For example, the
formation of polymers at negative potentials was improved with mediators [140].
The range of reductive redox mediators offers a powerful toolbox in cathodic organic
electrosyntheses with examples such as Fe(II) [141], Co(II)salen [142], Co(II) in vitamin B$_{12}$ [143], Ni(II) complexes, or Ti(IV) complexes [144], but also nonmetallic
aromatics such as naphthalene-, fluorene- [145], or perylene-derivatives [146,147].
The application of palladium complexes as cathodic redox mediators in the "electrochemically assisted" Heck reaction [148] represents an example where reaction rates
can be dramatically increased by passing current through a reaction while maintaining mild reaction conditions. Moeller and coworkers employed an undivided cell
with a battery power source to demonstrate a diverse range of the Heck coupling
chemistry (Scheme 1.8) for a range of systems and derivatives. The underlying mechanism is suggested to be based on the continuous regeneration of Pd(0) reaction
intermediates that then undergoes oxidative addition with the iodoaromatic substrate
before transferring the aromatic ring to the olefin.

The concept of "cascade reactions" has been recently highlighted in a review by
Elinson et al. [149]. Several examples are provided for both anodic and cathodic cascades in which multiple reactions occur sequentially triggered by electron transfer. The
cathodic formation of an indole at a palladium electrode [150] is an example of cyclization followed by reduction and electrophilic attack to give a range of indole products.
The process occurs in a simple undivided cell under galvanostatic control (Scheme 1.9).

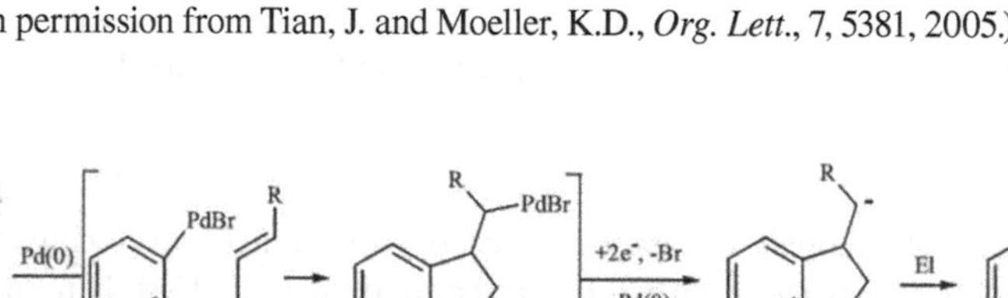

SCHEME 1.8 Reaction scheme for the electro-assisted Heck reaction. RVC, reticulated vitreous
carbon. (With permission from Tian, J. and Moeller, K.D., *Org. Lett.*, 7, 5381, 2005.)

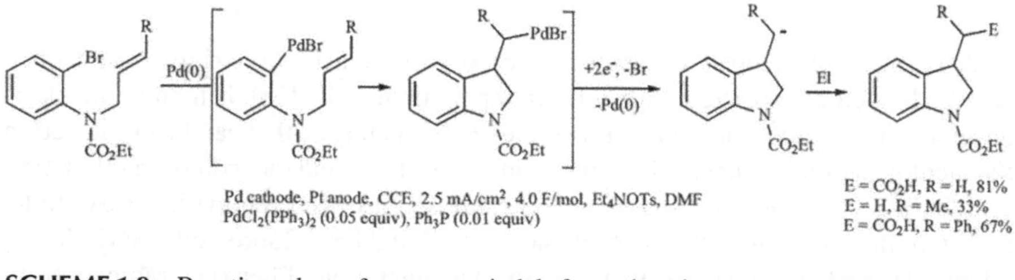

SCHEME 1.9 Reaction scheme for two-step indole formation via organometallic intermediates.
(With permission from Tanaka, H. et al., *Novel Trends in Electroorganic Synthesis*, Kodansha,
Tokyo, 1995.)

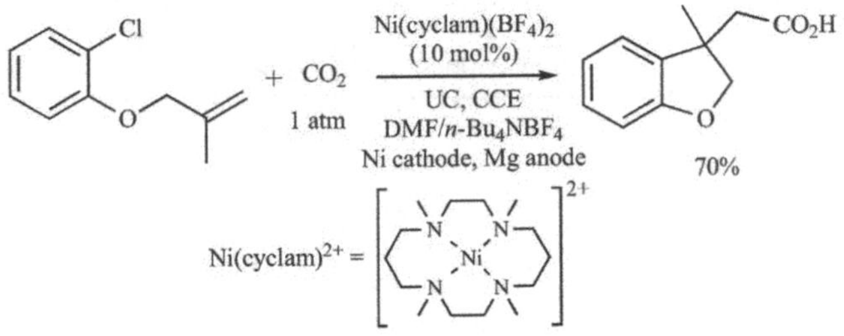

SCHEME 1.10 Reaction scheme for a Ni cathode-driven cyclization with CO_2 addition. (With permission from Olivero, S. and Dunach, E., *Eur. J. Org. Chem.*, 1999, 1885, 1999.)

A related example based on Ni(II)cyclam redox mediator in a cathodic cascade process is shown in Scheme 1.10 [151]. This process occurs in an undivided cell in DMF and at a nickel cathode. Carbon dioxide is employed as an electrophile to result in carboxylic-acid-functionalized products. The magnesium anode is sacrificial and not interfering in the synthesis.

These examples provide evidence for the powerful methodology that is available even with relatively simple/practical methods based on undivided reactor cells and constant current power supplies. However, there is usually only one of the two electrodes (anode and cathode) usefully employed and sacrificial or wasteful processes occur at the second. Therefore, further effort is needed to explore "paired electrosynthesis" where both anode and cathode processes are used to generated products.

1.4 METHODOLOGY FOR PAIRED TRANSFORMATIONS

Pairing electroorganic processes refers to the combining of anodic and cathodic processes [152]. Both are obviously needed and often sacrificial reactions or hydrogen evolution side reactions are tolerated in order to generate the desired product. However, when performing electrolysis reactions in the presence of both anode and cathode, reaction mechanisms can become rather complex and unpredictable. Very interesting pyridine products are generated, for example, during undivided cell electrolysis of acetonitrile at platinum electrodes [153] (see Scheme 1.11). When this reaction is performed in the presence of benzyl substrates, a new range of pyridine products are accessible [154]. Multiple reaction steps are needed, and a hypothetical reaction mechanism has been suggested.

A further example of a paired co-electrolysis has been developed by Elinson and coworkers [155,156]. Scheme 1.12 shows this process providing pyrimidine derivatives via spiro-cyclopropyl-barbiturate derivative intermediates under constant current electrolysis conditions in methanol and using NaBr as an active electrolyte. The electrode reactions here involve (i) methanol deprotonation at the cathode to give methoxy base (and hydrogen), which in turn deprotonates the barbiturate precursor, and (ii) bromine formation at the anode, which initiates cyclopropane formation. This reaction provides good yields in a relatively simple process employing both anode and cathode.

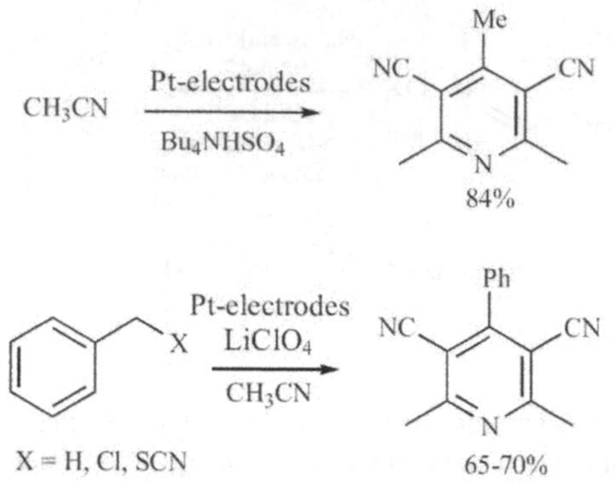

SCHEME 1.11 Paired electrolysis processes for the formation of pyridine derivatives at platinum electrodes under constant current conditions in acetonitrile. (With permission from Batanero, B. et al., *J. Org. Chem.*, 67, 2369, 2002; Otero, M.D. et al., *Tetrahedron Lett.*, 46, 8681, 2005.)

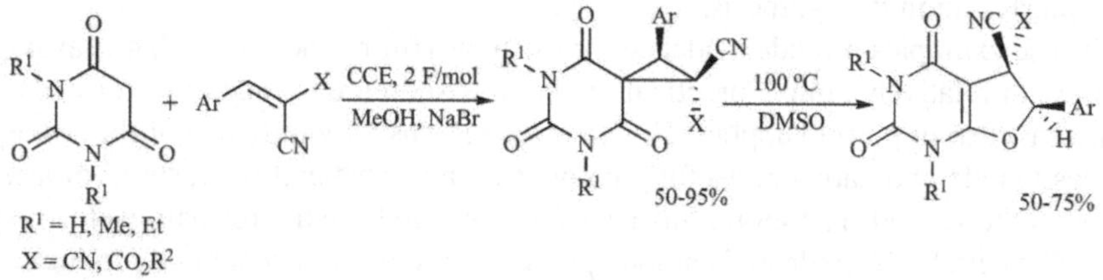

SCHEME 1.12 Paired electrolysis for the formation of pyrimidine derivatives. (With permission from Dorofeeva, E.O. et al., *RSC Adv.*, 2, 4444, 2012; Vereshchagin, A.N. et al., *RSC Adv.*, 5, 94986, 2015.)

Paired processes are mostly employed as spatially separated reactions either convergent (giving one product) or divergent or non-convergent (giving two separate products [157]). Convergent processes are particularly highlighted in a recent review by Baran [5]. An illustrative example for the convergent process is the Diels–Alder coupling of benzoquinone (produced at the anode) with an ortho-quinodimethane-diene (produced at the cathode) developed by Habibi et al. [158]. The process was performed in LiClO$_4$/ethanol and employing an undivided cell with galvanostatic control. Anode and cathode were graphite and Pb, respectively, and yields were scaled up and optimized to 57% (Scheme 1.13).

A further example of a convergent paired electrolysis (shown in Scheme 1.14) has been developed by Ishifune and coworkers [159]. An ester is reduced to alkoxide in THF solvent, and THF is oxidized at the α-carbon. The alkoxy-furan coupling products are obtained in good yield employing THF/'BuOH/LiClO$_4$ and with constant current applied to magnesium cathode and platinum anode.

Paired electrode processes in microgap microfluidic systems have been proposed [160] and offer the advantage of inter-diffusion of reaction intermediates from anode and cathode. This may complicate the overall reaction pathway, but also offers new opportunities for products from coupled reactions and cascades. He and coworkers

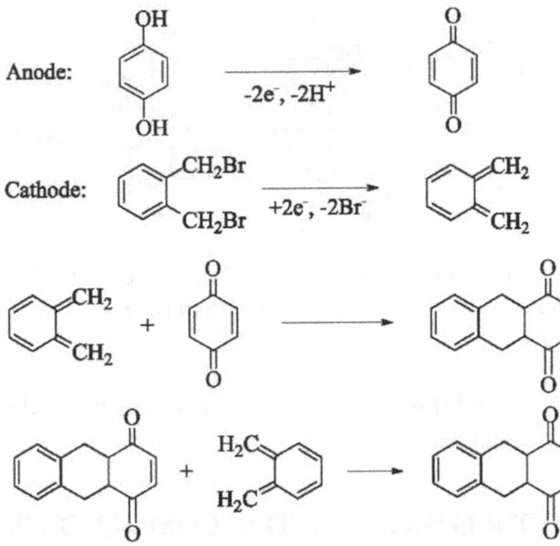

Anode:

OH → O (quinone), $-2e^-, -2H^+$

Cathode:

CH$_2$Br / CH$_2$Br → CH$_2$ / CH$_2$, $+2e^-, -2Br^-$

SCHEME 1.13 Convergent paired electrolysis of hydroquinone (anode) and 1,2-dibrommomethyl-benzene (cathode) producing polycyclic ketones via the Diels–Alder coupling. (With permission from Habibi, D. et al., *Electrochem. Commun.*, 49, 65, 2014.)

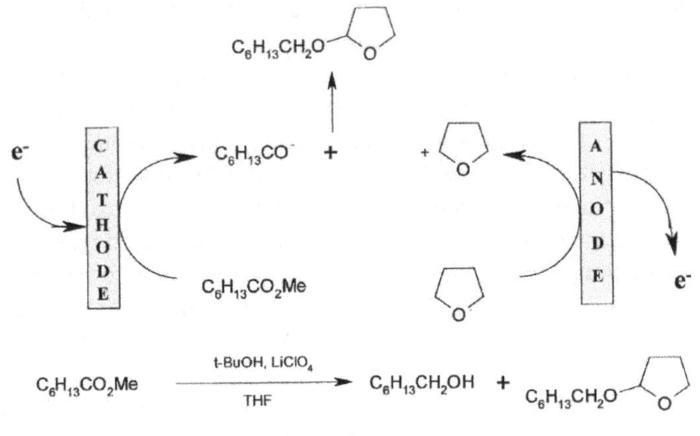

$C_6H_{13}CH_2O$—(furan ring)

e^- → CATHODE → $C_6H_{13}CO^-$ +

$C_6H_{13}CO_2Me$

ANODE → e^-

$$C_6H_{13}CO_2Me \xrightarrow[\text{THF}]{\text{t-BuOH, LiClO}_4} C_6H_{13}CH_2OH + C_6H_{13}CH_2O\text{—(furan)}$$

SCHEME 1.14 Paired convergent electrosynthesis of alkoxy-furans based on oxidation of THF and reduction of an ester precursor. (With permission from Ishifune, M. et al., *Electrochim. Acta*, 46, 20, 2001.)

used platinum–platinum microflow cells (Figure 1.25) and solutions of maleate esters and benzylbromide in DMF without intentionally added electrolyte to generate products in good yield under optimized flow and constant current conditions.

There are many more examples of paired electrolysis processes, but generally, much more work is needed to make the underlying reaction types more systematic and to allow a more rational approach to the planning and optimization of organic electrosyntheses under these conditions (possibly employing new computational tools). There is still a considerable level of adventure in designing new

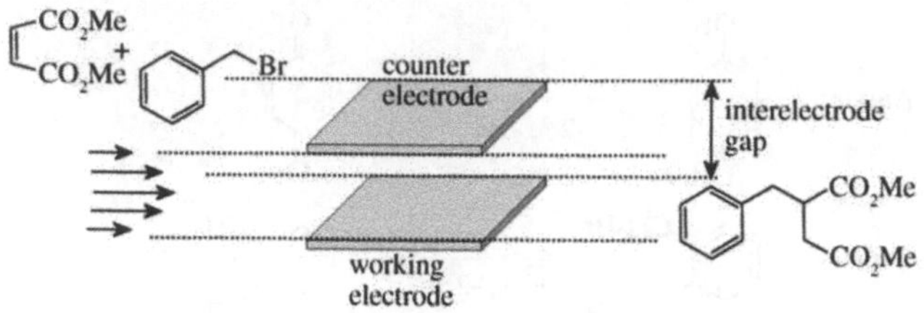

FIGURE 1.25 Drawing of a microfluidic platinum–platinum reactor cell with 45 mm² working area and 160 μm interelectrode gap. (With permission from He, P. et al., *Angew. Chem. Int. Ed.*, 45, 4146, 2006.)

electrosynthetic processes, and this is linked to opportunities in the design of novel and powerful synthetic strategies.

1.5 OUTLOOK, CONCLUSIONS, AND RECOMMENDATIONS

Methods for electroorganic synthesis are readily available, and there is considerable potential for de novo organic synthetic approaches based on electrochemically generated reactive intermediates and effective reactor systems. Interest in new electroorganic synthesis methodology is growing in both (i) the molecular-level understanding and exploitation of locally generated reactive intermediates and (ii) the reactor design and optimization of conditions.

It might be too much to ask for both clever new reaction pathways, utilization of oxidized and reduced intermediates, and at the same time optimized reactors and minimization of waste production and energy consumption. From the organic synthetic perspective, it is probably most important to allow for simplicity in the experimental approach (undivided cells, sacrificial counter electrodes, galvanostatic control with low cost power supplies, and a minimum of added reagents and salts). With more widely available and appreciated laboratory tools to perform electroorganic syntheses (see, e.g., [161]), there is likely to be a much stronger development in knowledge and uptake of knowledge into industrial contexts with further optimization in future.

Simplicity in the experimental approach is not necessarily served only in undivided cells and sacrificial electrodes. With a better understanding of the underlying chemistry, it is possible to engineer conditions for flow and microreactor conditions to optimize yields, to avoid salts and electrolytes, to minimize/recover energy usage, and to couple reactions beneficially. However, the approach of separately developing electrochemical processes and electrochemical reactors ignores an important element of adventure and opportunity for new chemistry when exploring microflow or microgap reactor systems: unexpected benefits may result from paired or coupled reactions. More work and synergy in chemical synthesis and chemical engineering approaches will be needed to provide a better foundation of electroorganic reaction types, as well as reactor types, with the aim to finally allow effective planning of electrosynthetic pathways in a way currently only possible in classical organic synthesis aided by modern computational tools [162].

REFERENCES

1. Hammerich, O.; Speiser, B. 2015. *Organic Electrochemistry*, 5th ed. New York: CRS Press Taylor & Francis Group.
2. Grimshaw, J. 2000. *Electrochemical Reactions and Mechanisms in Organic Chemistry.* Amsterdam: Elsevier.
3. Fuchigami, T.; Atobe, M.; Inagi, S. 2015. *Fundamentals and Applications of Organic Electrochemistry: Synthesis, Materials, Devices.* Chichester, UK: John Wiley & Sons Ltd.
4. Lund, H. 2002. A century of organic electrochemistry. *J. Electrochem. Soc.* 149: S21–S33.
5. Yan, M.; Kawamata, Y.; Baran, P.S. 2017. Synthetic organic electrochemical methods since 2000: On the verge of a renaissance. *Chem. Rev.* 117: 13230–13319.
6. Wiebe, A.; Gieshoff, T.; Mçhle, S.; Rodrigo, E.; Zirbes, M.; Waldvogel, S.R. 2018. Electrifying organic synthesis. *Angew. Chem. Int. Ed.* 57: 5594–5619.
7. Frontana-Uribe, B.A.; Little, R.D.; Ibanez, J.G.; Palma, A.; Vasquez-Medrano, R. 2010. Organic electrosynthesis: A promising green methodology in organic chemistry. *Green Chem.* 12: 2099–2119.
8. Pletcher, D. 2018. Organic electrosynthesis: A road to greater application. A mini review. *Electrochem. Commun.* 88: 1–4.
9. Horn, E.J.; Rosen, B.R.; Baran, P.S. 2016. Synthetic organic electrochemistry: An enabling and innately sustainable method. *ACS Central Sci.* 2: 302–308.
10. Lanzafame, P.; Abate, S.; Ampelli, C.; Genovese, C.; Passalacqua, R.; Centi, G.; Perathoner, S. 2017. Beyond solar fuels: Renewable energy-driven chemistry. *ChemSusChem* 10: 4409–4419.
11. Tremblay, P.L.; Zhang, T. 2015. Electrifying microbes for the production of chemicals. *Front. Microbiol.* 6: UNSP 201.
12. Bajracharya, S.; Srikanth, S.; Mohanakrishna, G.; Zacharia, R.; Strik, D.P.B.T.B.; Pant, D. 2017. Biotransformation of carbon dioxide in bioelectrochemical systems: State of the art and future prospects. *J. Power Sources* 356: 256–273.
13. Moeller, K.D. 2000. Synthetic applications of anodic electrochemistry. *Tetrahedron* 56: 9527–9554.
14. Birch, A.J.; Hinde, A.L.; Radom, L. 1981. A theoretical approach to the Birch reduction: Structures and stabilities of cyclohexadienes. *J. Am. Chem. Soc.* 103: 284–289.
15. VanAndelscheffer, P.J.M.; Barendrecht, E. 1995. Review on the electrochemistry of solvated electrons: Its use in hydrogenation of monobenzeoids. *Recuil Travaux Chimiques Pays-Bas—J. Roy. Netherlands Chem. Soc.* 114: 259–265.
16. Benkeser, R.A.; Lambert, R.F.; Kaiser, E.M. 1964. Selective reduction of aromatic compounds to dihydro or tetrahydro products by electrochemical method. *J. Am. Chem. Soc.* 86: 5272–5273.
17. Francke, R.; Little, R.D. 2014. Redox catalysis in organic electrosynthesis: Basic principles and recent developments. *Chem. Soc. Rev.* 43: 2492–2521.
18. DeAzevedo, D.C.; Goulart, M.O.F. 1997. Stereoselectivity in electrodic reactions. *Quimica Nova* 20: 158–169.
19. Edinger, C.; Kulisch, J.; Waldvogel, S.R. 2015. Stereoselective cathodic synthesis of 8-substituted (1R,3R,4S)-menthylamines. *Beilstein J. Org. Chem.* 11: 294–301.
20. Yutthalekha, T.; Wattanakit, C.; Lapeyre, V.; Nokbin, S.; Warakulwit, C.; Limtrakul, J.; Kuhn, A. 2016. Asymmetric synthesis using chiral-encoded metal. *Nat. Commun.* 7: 12678.
21. Wattanakit, C.; Yutthalekha, T.; Asssavapanumat, S.; Lapeyre, V.; Kuhn, A. 2017. Pulsed electroconversion for highly selective enantiomer synthesis. *Nat. Commun.* 8: 2087.

22. Fukuyama, T.; Rahman, T.; Sato, M.; Ryu, I. 2008. Adventures in inner space: Microflow systems for practical organic synthesis. *Synlett* 2: 151–163.
23. Wouters, B.; Hereijgers, J.; De Malsche, W.; Breugelmans, T.; Hubin, A. 2016. Electrochemical characterisation of a microfluidic reactor for cogeneration of chemicals and electricity. *Electrochim. Acta* 210: 337–345.
24. Cardoso, D.S.P.; Sljukic, B.; Santos, D.M.F.; Sequeira, C.A.C. 2017. Organic electrosynthesis: From laboratorial practice to industrial applications. *Org. Proc. Res. Dev.* 21: 1213–1226.
25. Atobe, M. 2017. Organic electrosynthesis in flow microreactor. *Curr. Opin. Electrochem.* 2: 1–6.
26. Watts, K.; Baker, A.; Wirth, T. 2014. Electrochemical synthesis in microreactors. *J. Flow Chem.* 4: 2–11.
27. Wiles, C.; Watts, P. 2011. Recent advances in micro reaction technology. *Chem. Commun.* 47: 6512–6535.
28. Yoshida, J.; Suga, S.; Nagaki, A. 2005. Selective organic reactions using microreactors. *J. Synth. Org. Chem. Jpn.* 63: 511–522.
29. Yoshida, J.; Shimizu, A.; Hayashi, R. 2018. Electrogenerated cationic reactive intermediates: The pool method and further advances. *Jpn. Chem. Rev.* 118: 4702–4730.
30. Jones, A.M.; Banks, C.E. 2014. The Shono-type electroorganic oxidation of unfunctionalised amides. Carbon–carbon bond formation via electrogenerated *N*-acyliminium ions. *Beilstein J. Org. Chem.* 10: 3056–3072.
31. Poe, S.L.; Cummings, M.A.; Haaf, M.R.; McQuade, D.T. 2006. Solving the clogging problem: Precipitate-forming reactions in flow. *Angew. Chem. Int. Ed.* 45: 1544–1548.
32. Löwe, H.; Axinte, R. D.; Breuch, D.; Hofmann, C. 2009. Heat pipe controlled syntheses of ionic liquids in microstructured reactors. *Chem. Eng. J.* 155: 548–550.
33. Ammonite by Cambridge Reactor Design (CRD). http://www.cambridgereactordesign. com/ammonite/ammonite.html (last visited on May 4, 2018).
34. Folgueiras-Amador, A.A.; Qian, X.-Y.; Xu, H.-C.; Wirth, T. 2018. Catalyst- and supporting-electrolyte-free electrosynthesis of benzothiazoles and thiazolopyridines in continuous flow. *Chem. Eur. J.* 24: 487–491.
35. Horn, E.J.; Rosen, B.R.; Chen, Y.; Tang, J.Z.; Chen, K.; Eastgate, M.D.; Baran, P.S. 2016. Scalable and sustainable electrochemical allylic C-H oxidation. *Nature* 533: 77–81.
36. Watkins, J.D.; Taylor, J.E.; Bull, S.D.; Marken, F. 2012. Mechanistic aspects of aldehyde and imine electro-reduction in a liquid-liquid carbon nanofiber membrane microreactor. *Tetrahedron Lett.* 5: 3357–3360.
37. Wadhawan, J.D.; Marken, F.; Compton, R.G. 2001. Biphasic sonoelectrosynthesis. A Review. *Pure Appl. Chem.* 73: 1947–1955.
38. Pletcher, D.; Walsh, F.C. 1990. *Industrial Electrochemistry*. London: Springer Science & Business Media.
39. Bard, A.J.; Faulkner, L.R. 2001. *Electrochemical Methods*. New York: Wiley & Sons.
40. Ball, J.C.; Compton, R.G.; Brett, C.M.A. 1998. Theory of anodic stripping voltammetry at wall-jet electrodes. Simulation of spatially differential stripping and redeposition phenomena. *J. Phys. Chem. B* 102: 162–166.
41. Macpherson, J.V.; Simjee, N.; Unwin, P.R. 2001. Hydrodynamic ultramicroelectrodes: Kinetic and analytical applications. *Electrochim. Acta* 47: 29–45.
42. Cooper, J.A.; Compton, R.G. 1998. Channel electrodes: A review. *Electroanalysis* 10: 141–155.
43. Ahn, S.D.; Frith, P.E.; Fisher, A.C.; Bond, A.M.; Marken, F. 2014. Hydrodynamic voltammetry at a rocking disc electrode: Theory versus experiment. *J. Electroanal. Chem.* 722: 78–82.

44. Streeter, I.; Compton, R.G. 2007. Steady state voltammetry at non-uniformly accessible electrodes: A study of Tafel plots for microdisc and tubular flow electrodes in the reversible and irreversible limits of electron transfer. *Phys. Chem. Chem. Phys.* 9: 862–870.
45. Brett, C.M.A.; Brett, A.M.O. 1993. *Electrochemistry: Principles, Methods, and Applications.* Oxford: Oxford University Press.
46. Hotchen, C.E.; Nguyen, H.V.; Fisher, A.C.; Frith, P.E.; Marken, F. 2015. Hydrodynamic microgap voltammetry under Couette flow conditions: Electrochemistry at a rotating drum in viscous poly(ethylene glycol). *ChemPhysChem* 16: 2789–2796.
47. Buckingham, M.A.; Cunningham, W.; Bull, S.D.; Buchard, A.; Folli, A.; Murphy, D.M.; Marken, F. 2018. Electrochemically driven C–H hydrogen abstraction processes with the tetrachloro-phthalimido-N-oxyl (Cl4PINO) catalyst. *Electroanalysis* doi. org/10.1002/elan.201800147.
48. Gorgy, K.; Lepretre, J.C.; Saint-Aman, E.; Einhorn, C.; Einhorn, J.; Marcadal, C.; Pierre, J.L. 1998. Electrocatalytic oxidation of alcohols using substituted N-hydroxyphthalimides as catalysts. *Electrochim. Acta* 44: 385–393.
49. Recupero, F.; Punta, C. 2007. Free radical functionalization of organic compounds catalyzed by N-hydroxyphthalimide. *Chem. Rev.* 107: 3800–3842.
50. Sholl, D.S.; Steckel, J.A. 2009. *Density Functional Theory: A Practical Introduction.* New York: John Wiley & Sons, Inc.
51. Raynal, F.; Barhdadi, R.; Perichon, J.; Savall, A.; Troupel, M. 2002. Water as solvent for nickel-2,2'-bipyridine-catalysed electrosynthesis of biaryls from haloaryls. *Adv. Synth. Catal.* 344: 45–49.
52. Carrero, H.; Gao, J.X.; Rusling, J.F.; Lee, C.W.; Fry, A.J. 1999. Direct and catalyzed electrochemical syntheses in microemulsions. *Electrochim. Acta* 45: 503–512.
53. Yoshida, J.; Kataoka, K.; Horcajada, R.; Nagaki, A. 2008. Modern strategies in electro-organic synthesis. *Chem. Rev.* 108: 2265–2299.
54. Paddon, C.A.; Atobe, M.; Fuchigami, T.; He, P.; Watts, P.; Haswell, S.J.; Pritchard, G.J.; Bull, S.D.; Marken, F. 2006. Towards paired and coupled electrode reactions for clean organic microreactor electrosyntheses. *J. Appl. Electrochem.* 36: 617–634.
55. Reichardt, C. 2007. Solvents and solvent effects: An introduction. *Org. Proc. Res. Dev.* 11: 105–113.
56. Chum, H.L.; Koch, V.R.; Miller, L.L.; Osteryoung, R.A. 1975. Electrochemical scrutiny of organometallic iron complexes and hexamethylbenzene in a room temperature molten salt. *J. Am. Chem. Soc.* 97: 3264–3265.
57. Wilkes, J.S.; Zaworotko, M.J. 1992. Air and water stable 1-ethyl-3-methylimidazolium based ionic liquids. *J. Chem. Soc. Chem. Commun.* 965–967.
58. Plechkova, N.V.; Seddon, K.R. 2008. Applications of ionic liquids in the chemical industry. *Chem. Soc. Rev.* 37: 123–150.
59. Hapiot, P.; Lagrost, C. 2008. Electrochemical reactivity in room-temperature ionic liquids. *Chem. Rev.* 108: 2238–2264.
60. MacFarlane, D.R.; Seddon, K.R. 2007. Ionic liquids progress on the fundamental issues. *Aust. J. Chem.* 60: 3–5.
61. Bonhôte, P.; Dias, A.-P.; Papageorgiou, N.; Kalyanasundaram, K.; Grätzel, M. 1996. Hydrophobic, highly conductive ambient-temperature molten salts. *Inorg. Chem.* 35: 1168–1178.
62. McEwen, A.B.; Ngo, H.L.; LeCompte, K.; Goldman, J.L. 1999. Electrochemical properties of imidazolium salt electrolytes for electrochemical capacitor applications. *J. Electrochem. Soc.* 146: 1687–1695.
63. Sun, J.; Forsyth, M.; MacFarlane, D.R. 1998. Room-temperature molten salts based on the quaternary ammonium ion. *J. Phys. Chem. B* 102: 8858–8864.

64. Tokuda, H.; Hayamizu, K.; Ishii, K.; Susan, M.A.B.H.; Watanabe, M. 2005. Physicochemical properties and structures of room temperature ionic liquids. 2. Variation of alkyl chain length in imidazolium cation. *J. Phys. Chem. B* 109: 6103–6110.

65. Paiva, A.; Craveiro, R.; Aroso, I.; Martins, M.; Reis, R.L.; Duarte, A.R.C. 2014. Natural deep eutectic solvents–solvents for the 21st century. *ACS Sust. Chem. Eng.* 2: 1063–1071.

66. Suarez, P.A.Z.; Consorti, C.S.; Souza, R.F.; Dupont, J.; Gonçalves, R.S. 2002. Electrochemical behavior of vitreous glass carbon and platinum electrodes in the ionic liquid 1-*n*-butyl-3-methylimidazolium trifluoroacetate. *J. Braz. Chem. Soc.* 13: 106–109.

67. Fitchett, B.D.; Knepp, T.N.; Conboy, J.C. 2004. 1-Alkyl-3-methylimidazolium bis(perfluoroalkylsulfonyl)imide water-immiscible ionic liquids: The effect of water on electrochemical and physical properties. *J. Electrochem. Soc.* 151: E219–E225.

68. Schröder, U.; Wadhawan, J.D.; Compton, R.G.; Marken, F.; Suarez, P.A.Z.; Consorti, C.S.; de Souza, R.F.; Dupont, J. 2000. Water-induced accelerated ion diffusion: Voltammetric studies in 1-methyl-3-[2,6-(S)-dimethylocten-2-yl]imidazolium tetrafluoroborate, 1-butyl-3-methylimidazolium tetrafluoroborate and hexafluorophosphate ionic liquids. *New J. Chem.* 24: 1009–1015.

69. Zhang, S.; Sun, N.; He, X.; Lu, X.; Zhang, X. 2006. Physical properties of ionic liquids: Database and evaluation. *J. Phys. Chem. Ref. Data* 35: 1475–1517.

70. Hasegawa, M.; Ishii, H.; Fuchigami, T. 2003. Selective anodic fluorination of phthalides in ionic liquids. *Green Chem.* 5: 512–515.

71. Hu, Y.F.; Liu, Z.C.; Xu, C.M.; Zhang, X.M. 2011. The molecular characteristics dominating the solubility of gases in ionic liquids. *Chem. Soc. Rev.* 40: 3802–3823.

72. Lim, H.K.; Kim, H. 2017. The mechanism of room-temperature ionic-liquid-based electrochemical CO_2 reduction: A review. *Molecules* 22: 536.

73. Navarro, M. 2017. Recent advances in experimental procedures for electroorganic synthesis. *Curr. Opin. Electrochem.* 2: 43–52.

74. Zhao, S.F.; Horne, M.; Bond, A.M.; Zhang, J. 2016. Is the imidazolium cation a unique promoter for electrocatalytic reduction of carbon dioxide? *J. Phys. Chem. C* 120: 23989–24001.

75. Zhang, M.; Desai, T.; Ferrari, M. 1998. Proteins and cells on PEG immobilized silicon surfaces. *Biomaterials* 19: 953–960.

76. Downard, A.J.; bin Mohamed, A. 1999. Suppression of protein adsorption at glassy carbon electrodes covalently modified with tetraethylene glycol diamine. *Electroanalysis* 11: 418–423.

77. Maeda, H.; Itami, M.; Katayama, K.; Yamauchi, Y.; Ohmori, H. 1997. Anodization of glassy carbon electrodes in oligomers of ethylene glycol and their monomethyl ethers as a tool for the elimination of protein adsorption. *Anal. Sci.* 13: 721–727.

78. Yang, Z.-Z.; He, L.-N.; Zhao, Y.-N.; Li, B.; Yu, B. 2011. CO_2 capture and activation by superbase/polyethylene glycol and its subsequent conversion. *Energy Environ. Sci.* 4: 3971–3975.

79. Chen, L.; Li, F.W.; Zhang, Y.; Bentley, C.L.; Horne, M.; Bond, A.M.; Zhang, J. 2017. Electrochemical reduction of carbon dioxide in a monoethanolamine capture medium. *ChemSusChem* 10: 4109–4118.

80. McNicholas, B.J.; Blakemore, J.D.; Chang, A.B.; Bates, C.M.; Kramer, W.W.; Grubbs, R.H.; Gray, H.B. 2016. Electrocatalysis of CO_2 reduction in brush polymer ion gels. *J. Am. Chem. Soc.* 138: 11160–11163.

81. Kolbe, H. 1849. Untersuchungen über die elektrolyse organischer verbindungen. *Justus Liebigs Ann. Chem.*, 69: 257–294.

82. Stang, C.; Harnisch, F. 2016. The dilemma of supporting electrolytes for electroorganic synthesis: A case study on Kolbe electrolysis. *ChemSusChem* 9: 50–60.

83. Kurihara, H.; Fuchigami, T.; Tajima, T. 2008. Kolbe carbon–carbon coupling electrosynthesis using solid-supported bases. *J. Org. Chem.* 73: 6888–6890.

84. Zhang, L.; Zhang, Z.; Hong, J.; Yu, J.; Zhang, J.; Mo, F. 2018. Oxidant-free C(sp^2)-H functionalization/C–O bond formation: A Kolbe oxidative cyclization process. *J. Org. Chem.* 83: 3200–3207.

85. Zhang, S.; Li, L.J.; Wang, H.Q.; Li, Q.; Liu, W.M.; Xu, K.; Zeng, C.C. 2018. Scalable electrochemical dehydrogenative lactonization of C(sp^2/sp^3)-H bonds. *Org. Lett.* 20: 252–255.

86. Frey, D.A.; Wu, N.; Moeller, K.D. 1996. Anodic electrochemistry and the use of a 6-volt lantern battery: A simple method for attempting electrochemically based synthetic transformations. *Tetrahedron Lett.* 37: 8317–8320.

87. Nguyen, B.H.; Redden, A.; Moeller, K.D. 2014. Sunlight, electrochemistry, and sustainable oxidation reactions. *Green Chem.* 16: 69–72.

88. Xiao, Q.; Liu, Z.; Bo, A.; Zavahir, S.; Sarina, S.; Bottle, S.; Riches, J.D.; Zhu, H.Y. 2015. Catalytic transformation of aliphatic alcohols to corresponding esters in O$_2$ under neutral conditions using visible-light irradiation. *J. Am. Chem. Soc.* 137: 1956–1966.

89. Hao, H.C.; Zhang, L.; Wang, W.Z.; Zeng, S.W. 2018. Modification of heterogeneous photocatalysts for selective organic synthesis. *Catal. Sci. Technol.* 8: 1229–1250.

90. Friedmann, D.; Hakki, A.; Kim, H.; Choic, W.; Bahnemannd, D. 2016. Heterogeneous photocatalytic organic synthesis: State-of-the-art and future perspectives. *Green Chem.* 18: 5391–5411.

91. Stephenson, C.R.J.; Yoon, T.P.; MacMillan, D.W.C. (eds.) 2018. *Visible Light Photocatalysis in Organic Chemistry*. Weinheim, Germany: Wiley-VCH.

92. Tomita, O.; Ohtani, B.; Abe, R. 2014. Highly selective phenol production from benzene on a platinum-loaded tungsten oxide photocatalyst with water and molecular oxygen: Selective oxidation of water by holes for generating hydroxyl radical as the predominant source of the hydroxyl group. *Catal. Sci. Technol.* 4: 3850–3860.

93. Whitehead, C.R.; Sessions, E.H.; Ghiviriga, I.; Wright, D.L. 2002. Two-step electrochemical annulation for the assembly of polycyclic systems. *Org. Lett.* 4: 3763–3765.

94. Sperry, J.B.; Whitehead, C.R.; Ghiviriga, I.; Walczak, R.M.; Wright, D.L. 2004. Electrooxidative coupling of furans and silyl enol ethers: Application to the synthesis of annulated furans. *J. Org. Chem.* 69: 3726–3734.

95. Sperry, J.B.; Wright, D.L. 2005. The gem-dialkyl effect in electron transfer reactions: Rapid synthesis of seven-membered rings through an electrochemical annulation. *J. Am. Chem. Soc.* 127: 8034–8035.

96. Sperry, J.B.; Wright, D.L. 2006. Annulated heterocycles through a radical-cation cyclization: Synthetic and mechanistic studies. *Tetrahedron* 62: 6551–6557.

97. Sperry, J.B.; Wright, D.L. 2005. Synthesis of the hamigeran skeleton through an electro-oxidative coupling reaction. *Tetrahedron Lett.* 46: 411–414.

98. Wu, J.W.; Zhou, Y.; Zhou, Y.C.; Chiang, C.W.; Lei, A.W. 2017. Electro-oxidative C(sp^3)-H amination of azoles via intermolecular oxidative C(sp^3)-H/N-H cross-coupling. *ACS Catal.* 7: 8320–8323.

99. Kabeshov, M.A.; Musio, B.; Ley, S.V. 2017. Continuous direct anodic flow oxidation of aromatic hydrocarbons to benzyl amides. *React. Chem. Eng.* 2: 822–825.

100. O'Brien, A.G.; Maruyama, A.; Inokuma, Y.; Fujita, M.; Baran, P.S.; Blackmond, D.G. 2014. Radical C-H functionalization of heteroarenes under electrochemical control. *Angew. Chem. Int. Ed.* 53: 11868–11871.

101. Frankowski, K.J.; Liu, R.Z.; Milligan, G.L.; Moeller, K.D.; Aubé, J. 2015. Practical electrochemical anodic oxidation of polycyclic lactams for late stage functionalization. *Angew. Chem. Int. Ed.* 54: 10555–10558.

102. Xiong, P.; Xu, H.-H.; Xu, H.-C. 2017. Metal- and reagent-free intramolecular oxidative amination of tri- and tetrasubstituted alkenes. *J. Am. Chem. Soc.* 139: 2956–2959.

103. Xu, F.; Qian, X.Y.; Li, Y.J.; Xu, H.C. 2017. Synthesis of 4H-1,3-benzoxazines via metal- and oxidizing reagent free aromatic C-H oxygenation. *Org. Lett.* 19: 6332–6335.

104. Imada, Y.; Okada, Y.; Chiba, K. 2018. Investigating radical cation chain processes in the electrocatalytic Diels-Alder reaction. *Beilstein J. Org. Chem.* 14: 642–647.

105. Imada, Y.; Yamaguchi, Y.; Shida, N.; Okada, Y.; Chiba, K. 2017. Entropic electrolytes for anodic cycloadditions of unactivated alkene nucleophiles. *Chem. Commun.* 53: 3960–3963.

106. Zhao, Y.; Lai, Y.L.; Du, K.S.; Lin, D.Z.; Huang, J.M. 2017. Electrochemical decarboxylative sulfonylation of cinnamic acids with aromatic sulfonylhydrazides to vinyl sulfones. *J. Org. Chem.* 82: 9655–9661.

107. Gao, Y.Y.; Wang, Y.; Zhou, J.; Mei, H.B.; Han, J.L. 2018. An electrochemical oxidative homo-coupling reaction of imidazopyridine heterocycles to biheteroaryls. *Green Chem.* 20: 583–587.

108. Andrieux, C.P.; Grzeszczuk, M.; Saveant, J.M. 1991. Electrochemical generation and reduction of organic free-radicals: Alpha-hydroxybenzyl radicals from the reduction of benzaldehyde. *J. Am. Chem. Soc.* 113: 8811–8817.

109. Libot, C.; Pletcher, D. 2000. The reduction of carbonyl compounds at carbon electrodes in acidic water/methanol mixtures. *Electrochem. Commun.* 2: 141–144.

110. Gütz, C.; Selt, M.; Bänziger, M.; Bucher, C.; Römelt, C.; Hecken, N.; Gallou, F.; Galvão, T.R.; Waldvogel, S.R. 2015. A novel cathode material for cathodic dehalogenation of 1,1-dibromo cyclopropane derivatives. *Chem. Eur. J.* 21: 13878–13882.

111. Gütz, C.; Bänziger, M.; Bucher, C.; Galvão, T.R.; Waldvogel, S.R. 2015. Development and scale-up of the electrochemical dehalogenation for the synthesis of a key intermediate for NS5A inhibitors. *Org. Proc. Res. Dev.* 19: 1428–1433.

112. Birch, A.J. 1946. Electrolytic reduction in liquid ammonia. *Nature* 158: 60–61.

113. Hook, J.M.; Mandler, L.N. 1986. Recent developments in the Birch reduction of aromatic compounds: Applications to the synthesis of natural products. *Nat. Prod. Rep.* 3: 35–85.

114. Sternberg, H.W.; Markby, R.E.; Wender, I.; Mohliner, D.M. 1966. Electrochemical reduction of aromatic hydrocarbons in ethylenediamine. *J. Electrochem. Soc.* 113: 1060–1061.

115. Del Campo, F.J.; Neudeck, A.; Compton, R.G.; Marken, F.; Bull, S.D.; Davies, S.G. 2001. Low-temperature sonoelectrochemical processes part 2: Generation of solvated electrons and Birch reduction processes under high mass transport conditions in liquid ammonia. *J. Electroanal. Chem.* 507: 144–151.

116. Senboku, H.; Katayama, A. 2017. Electrochemical carboxylation with carbon dioxide. *Curr. Opin. Green Sust. Chem.* 3: 50–54.

117. Matthessen, R.; Fransaer, J.; Binnemans, K.; De Vos, D.E. 2014. Electrocarboxylation: Towards sustainable and efficient synthesis of valuable carboxylic acids. *Beilstein J. Org. Chem.* 10: 2484–2500.

118. Qu, Y.; Tsuneishi, C.; Tateno, H.; Matsumurab, Y.; Atobe, M. 2017. Green synthesis of alpha-amino acids by electrochemical carboxylation of imines in a flow microreactor. *React. Chem. Eng.* 2: 871–875.

119. Luo, P.P.; Zhang, Y.T.; Chen, B.L.; Yu, S.X.; Zhou, H.W.; Qu, K.G.; Kong, Y.X.; Huang, X.Q.; Zhang, X.X.; Lu, J.X. 2017. Electrocarboxylation of dichlorobenzenes on a silver electrode in DMF. *Catalysis* 7: 274.

120. Raju, T.; Basha, C.A. 2005. Electrochemical cell design and development for mediated electrochemical oxidation-Ce(III)/Ce(IV) system. *Chem. Eng. J.* 114: 55–65.
121. Steckhan, E. 1986. Indirect electroorganic synthesis: A modern chapter of organic electrochemistry. *Angew. Chem. Int. Ed.* 25: 683–701.
122. Kolodziej, A.; Ahn, S.D.; Carta, M.; Malpass-Evans, R.; McKeown, N.B.; Chapman, R.S.L.; Bull, S.D.; Marken, F. 2015. Electrocatalytic carbohydrate oxidation with 4-benzoyloxy-TEMPO heterogenised in a polymer of intrinsic microporosity. *Electrochim. Acta* 160: 195–201.
123. Steckhan, E. 1987. Organic syntheses with electrochemically regenerable redox systems. E. Steckhan, Ed., In *Topics in Organic Chemistry*. Heidelberg, Berlin: Springer-Verlag.
124. Bartlett, P.N. 2008. *Bioelectrochemistry: Fundamentals, Experimental Techniques and Applications*. New York: John Wiley & Sons.
125. Kawamata, Y.; Yan, M.; Liu, Z.Q.; Bao, D.H.; Chen, J.S.; Starr, J.T.; Baran, P.S. 2017. Scalable, electrochemical oxidation of unactivated C-H bonds. *J. Am. Chem. Soc.* 139: 7448–7451.
126. Horn, E.J.; Rosen, B.R.; Chen, Y.; Tang, J.; Chen, K.; Eastgate, M.D.; Baran, P.S. 2016. Scalable and sustainable electrochemical allylic C-H oxidation. *Nature* 533: 77–81.
127. Shrestha, A.; Lee, M.; Dunn, A.L.; Sanford, M.S. 2018. Palladium-catalyzed C-H bond acetoxylation via electrochemical oxidation. *Org. Lett.* 20: 204–207.
128. Yang, Q.L.; Li, Y.Q.; Ma, C.; Fang, P.; Zhang, X.J.; Mei, T.S. 2017. Palladium-catalyzed C(sp³)-H oxygenation via electrochemical oxidation. *J. Am. Chem. Soc.* 139: 3293–3298.
129. Johnson, B.M.; Francke, R.; Little, R.D.; Berben, L.A. 2017. High turnover in electro-oxidation of alcohols and ethers with a glassy carbon-supported phenanthroimidazole mediator. *Chem. Sci.* 8: 6493–6498.
130. Broese, T.; Francke, R. 2016. Electrosynthesis using a recyclable mediator-electrolyte system based on ionically tagged phenyl iodide and 1,1,1,3,3,3-hexafluoroisopropanol. *Org. Lett.* 18: 5896–5899.
131. Jiang, Y.Y.; Wang, Q.Q.; Liang, S.; Hu, L.M.; Little, R.D.; Zeng, C.C. 2016. Electrochemical oxidative amination of sodium sulfinates: Synthesis of sulfonamides mediated by NH₄I as a redox catalyst. *J. Org. Chem.* 81: 4713–4719.
132. Bosque, I.; Magallanes, G.; Rigoulet, M.; Karkas, M.D.; Stephenson, C.R.J. 2017. Redox catalysis facilitates lignin depolymerization. *ACS Central Sci.* 3: 621–628.
133. Sauermann, N.; Meyer, T.H.; Tian, C.; Ackermann, L. 2017. Electrochemical cobalt-catalyzed C-H oxygenation at room temperature. *J. Am. Chem. Soc.* 139: 18452–18455.
134. Fu, N.; Sauer, G.S.; Lin, S. 2017. Electrocatalytic radical dichlorination of alkenes with nucleophilic chlorine sources. *J. Am. Chem. Soc.* 139: 15548–15553.
135. Ye, K.-Y.; Pombar, G.; Fu, N.; Sauer, G.S.; Keresztes, I.; Lin, S. 2018. Anodically coupled electrolysis for the heterodifunctionalization of alkenes. *J. Am. Chem. Soc.* 140: 2438–2441.
136. Hou, Z.-W.; Mao, Z.-Y.; Song, J.; Xu, H.-C. 2017. Electrochemical synthesis of polycyclic N-heteroaromatics through cascade radical cyclization of diynes. *ACS Catal.* 7: 5810–5813.
137. Xiong, P.; Xu, H.H.; Song, J.S.; Xu, H.C. 2018. Electrochemical difluoromethyl-arylation of alkynes. *J. Am. Chem. Soc.* 140: 2460–2464.
138. Lennox, A.J.J.; Nutting, J.E.; Stahl, S.S. 2018. Selective electrochemical generation of benzylic radicals enabled by ferrocene-based electron-transfer mediators. *Chem. Sci.* 9: 356–361.
139. Li, L.; Luo, S. 2018. Electrochemical generation of diaza-oxyallyl cation for cycloaddition in an all-green electrolytic system. *Org. Lett.* 20: 1324–1327.

140. Peres, L.O.; Utley, J.H.P.; Gruber, J. 2004. The advantageous use of mediators in the electrosynthesis of poly(p-phenylenevinylene)s. *Electrochem. Commun.* 6: 1141–1143.
141. Feng, D.M.; Zhu, Y.P.; Chen, P.; Ma, T.Y. 2017. Recent advances in transition-metal-mediated electrocatalytic CO_2 reduction: From homogeneous to heterogeneous systems. *Catalysis* 7: 373.
142. Chen, B.L.; Zhu, H.W.; Xiao, Y.; Sun, Q.L.; Wang, H.; Lu, J.X. 2014. Cosalen asymmetric electrocarboxylation of 1-phenylethyl chloride catalyzed by electrogenerated chiral [Co(I)(salen)]⁻ complex. *Electrochem. Commun.* 42: 55–59.
143. Rusling, J.F.; Zhou, D.L. 1997. Electrochemical catalysis in microemulsions. Dynamics and organic synthesis. *J. Electroanal. Chem.* 439: 89–96.
144. Torii, S. 1993. Organometal complexes in electroorganic synthesis. *J. Synth. Org. Chem. Japan* 51: 1024–1042.
145. Mitsudo, K.; Nakagawa, Y.; Mizukawa, J.; Tanaka, H.; Akaba, R.; Okada, T.; Suga, S. 2012. Electro-reductive cyclization of aryl halides promoted by fluorene derivatives. *Electrochim. Acta* 82: 444–449.
146. Koshechko, V.G.; Pokhodenko, V.D. 2001. Electrochemical activation of freons using electron transfer mediators. *Russ. Chem. Bull.* 50: 1929–1935.
147. Sun, G.Q.; Ren, S.Y.; Zhu, X.H.; Huang, M.N.; Wan, Y.Q. 2016. Direct arylation of pyrroles via indirect electroreductive C-H functionalization using perylene bisimide as an electron-transfer mediator. *Org. Lett.* 18: 544–547.
148. Tian, J.; Moeller, K.D. 2005. Electrochemically assisted Heck reactions. *Org. Lett.* 7: 5381–5383.
149. Elinson, M.N.; Vereshchagin, A.N.; Ryzkov, F.V. 2017. Electrochemical synthesis of heterocycles via cascade reactions. *Curr. Org. Chem.* 21: 1427–1439.
150. Tanaka, H.; Ren, Q.; Torii, S. 1995. S. Torii, Ed., In *Novel Trends in Electroorganic Synthesis*, Modification of Cephalosporins Bearing Substituents at C(3')Position via Electrogenerated Radical-Cation of Sulfide 195–196, Tokyo: Kodansha.
151. Olivero, S.; Dunach, E. 1999. Selectivity in the tandem cyclization: Carboxylation reaction of unsaturated haloaryl ethers catalyzed by electrogenerated nickel complexes. *Eur. J. Org. Chem.* 1999: 1885–1891.
152. Li, W.; Nonaka, T.; Chou, T.C. 1999. Paired electrosynthesis of organic compounds. *Electrochemistry* 67: 4–10.
153. Batanero, B.; Barba, F.; Avelino, M. 2002. Preparation of 2,6-dimethyl-4-arylpyridine-3,5-dicarbonitrile: A paired electrosynthesis. *J. Org. Chem.* 67: 2369–2371.
154. Otero, M.D.; Batanero, B.; Barba, F. 2005. Electrosynthesis of pyridines from 'only acetonitrile'. *Tetrahedron Lett.* 46: 8681–8683.
155. Dorofeeva, E.O.; Elinson, M.N.; Vereshchagin, N.A.; Stepanov, N.O.; Bushmarinov, I.S.; Belyakov, P.A.; Sokolova, O.A.; Nikishin, G.I. 2012. Electrocatalysis in MIRC reaction strategy: Facile stereoselective approach to medicinally relevant spirocyclopropylbarbiturates from barbituric acids and activated olefins. *RSC Adv.* 2: 4444–4452.
156. Vereshchagin, A.N.; Elinson, M.N.; Dorofeeva, E.O.; Sokolova, O.O.; Bushmarinov, I.S.; Egorov, M.P. 2015. Stereoselective synthesis of medicinally relevant furo[2,3-d]pyrimidine framework by thermal rearrangement of spirocyclic barbiturates. *RSC Adv.*, 5: 94986–94989.
157. Babu, K.F.; Sivasubramanian, R.; Noel, M.; Kulandainathan, M.A. 2011. A homogeneous redox catalytic process for the paired synthesis of L-cysteine and L-cysteic acid from L-cystine. *Electrochim. Acta* 56: 9797–9801.
158. Habibi, D.; Pakravan, N.; Nematollahi, D. 2014. The green and convergent paired Diels-Alder electro-synthetic reaction of 1,4-hydroquinone with 1,2-bis(bromo- methyl)benzene. *Electrochem. Commun.* 49: 65–69.

159. Ishifune, M.; Yamashita, H.; Matsuda, M.; Ishida, H.; Yamashita, N.; Kera, Y.; Kashimura, S.; Masuda, H.; Murase, H. 2001. Electro reduction of aliphatic esters using new paired electrolysis systems. *Electrochim. Acta* 46: 20–21.
160. He, P.; Watts, P.; Marken, F.; Haswell, S.J. 2006. Self-supported and clean one-step cathodic coupling of activated olefins with benzyl bromide derivatives in a micro flow reactor. *Angew. Chem. Int. Ed.* 45: 4146–4149.
161. Yan, M.; Kawamata, Y.; Baran, P.S. 2018. Synthetic organic electrochemistry: Calling all engineers. *Angew. Chem. Int. Ed.* 57: 4149–4155.
162. Szymkuc, S.; Gajewska, E.P.; Klucznik, T.; Molga, K.; Dittwald, P.; Startek, M.; Bajczyk, M.; Grzybowski, B.A. 2016. Computer-assisted synthetic planning: The end of the beginning. *Angew. Chem. Int. Ed.* 55: 5904–5937.

2 Cogeneration Electrosynthesis

Xiangming Gao and Adrian C. Fisher
University of Cambridge

Nathan Lawrence
University of Hull

Peng Song
Beijing University of Technology

CONTENTS

2.1 Introduction 45
2.2 Electricity and Chemical Cogeneration Processes 48
    2.2.1 Introduction to Cogeneration 48
    2.2.2 Chemical Cogeneration in Aqueous Electrolyte Fuel Cells 49
    2.2.3 Chemical Cogeneration in Aqueous Phosphorous Acid Fuel Cells.... 51
    2.2.4 Chemical Cogeneration in Polymer Electrolyte Fuel Cells 52
    2.2.5 Chemical Cogeneration in Solid Oxide Fuel Cells 57
    2.2.6 Chemical Cogeneration in Molten Salt Fuel Cells 62
2.3 Conclusion 62
References 63

2.1 INTRODUCTION

Fuel cells are energy conversion devices that convert chemical energy directly into electric power. The efficiency of energy conversion through "thermal-mechanical-electric" sequence in conventional heat engines, e.g., steam turbines, is limited by Carnot's theorem limitation. Typically, a steam turbine power plant can only obtain an energy conversion efficiency of ca. 35% [1]. However, with the direct chemical energy-electricity conversion, fuel cells eliminate the limitations in the conventional indirect conversion process, allowing the theoretical conversion efficiency to reach almost 100%. The conversion efficiencies that have been achieved by gaseous or liquid fuel cells are between 40% and 60%, while even higher efficiencies, ca. 80%, can be achieved by direct carbon fuel cells, which are twice that of a steam turbine power plant [2]. Environmental factors make fuel cells receive more attention recently. Fuels and oxidants are supplied to fuel cells and react inside of the devices separately; thus, it is much easier to deal with the emissions, since most of the work of separation can

45

be eliminated. The higher conversion efficiency also leads to fewer emissions. Thus, this technology is much more environmentally benign than heat engines.

The first fuel cell devices were independently developed by William Grove and Christian Friedrich Schönbein almost simultaneously in 1838. However, the first commercially used fuel cells were not produced come until Apollo space program of the National Aeronautics and Space Administration in the 1960s. Since then, various modes of fuel cell systems have been developed. Fuel cells are frequently classified by the electrolytes employed. Another classification method is based on their operating temperatures, i.e., low-temperature fuel cells and high-temperature fuel cells. The subset of low-temperature fuel cells includes alkaline fuel cell (AFC), proton exchange membrane fuel cell (PEMFC), direct methanol fuel cells (DMFCs), phosphoric acid fuel cell (PAFC), etc. The subset of high-temperature fuel cells, which are normally operated at 500–1000°C, consists of two different types, i.e., the molten carbonate fuel cell (MCFC) and the solid oxide fuel cell (SOFC). Table 2.1 lists fuel cells classified by their electrolytes and the characteristics of each type.

Generally, fuel cells are designed as a chemical-electric energy converter. However, they can also be operated as an electrochemical reactor, depending on which mode they are operated in. Three typical operating modes of fuel cell systems are given: (1) electric energy generation, (2) electric energy and useful chemical cogeneration, and (3) electrochemical "pumping" mode [4]. The process of how the three modes work is shown in Figure 2.1. Except mode (3), the other two modes both proceed spontaneously, while external power source is employed in mode (3) to drive a current through the cell in the desired direction, and only chemicals are generated. In this review, however, only mode (2), i.e., electric energy and useful chemical cogeneration, will be discussed. The main advantages seen in this technology are the high recovering efficiency of chemical energy into the form of electricity, avoiding efficiency loss caused by Carnot thermal cycles, and the natural separation of products in cathode and anode chambers. The latest development of cogeneration

TABLE 2.1
Types of Fuel Cells Classified by the Electrolytes Employed and Their Characteristics

	AFC	PEMFC	DMFC	PAFC	MCFC	SOFC
Operating temperature	<100°C	85–105°C	85–105°C	150–200°C	600–700°C	800–1000°C
Electrode material	Pt/C gas diffusion electrode	Pt-on-carbon	Pt-on-carbon	Pt-on-carbon	Ni+Cr	Ni/gallium-doped cerium oxide, LSM/YSZ
Electrolyte	KOH	Polymer membrane	Polymer membrane	H_3PO_4	Molten carbonate	YSZ
Fuel	H_2	H_2	Methanol	H_2 reformate	H_2/CO reformate	$H_2/CO/CH_4$ reformate
Oxidant	O_2/air	O_2/air	O_2	O_2/air	CO_2/O_2/air	O_2/air

Source: With permission from Carrette, L. et al., *ChemPhysChem*, 1, 162, 2000.

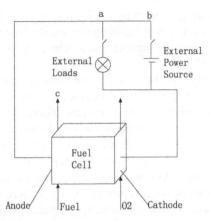

FIGURE 2.1 Configuration of a fuel cell operated in different modes: 1—(a) access works, electric energy generation; 2—(a) and (c) accesses work, electricity and chemical cogeneration; 3—(b) and (c) accesses work, electrochemical "pumping" mode.

technology will be discussed subsequently, while previous development in this area has been reviewed by Alcaide et al. [5].

Figure 2.2 shows the schematic of an electricity and chemical cogeneration system, including the chemical flows and the electron flows involved in the process.

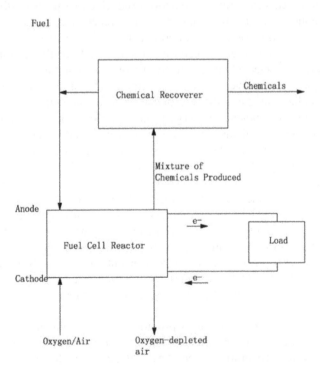

FIGURE 2.2 Schematic of electricity and chemical cogeneration system.

Typically, a cogeneration system consists of a prototypical fuel cell that generates electricity and chemicals, external loads that consume the electric power generated by the fuel cell, and a chemical recoverer that separates the useful chemicals from the mixture of the product.

A variety of electricity and chemical cogeneration systems have been developed. Many chemicals obtained from these processes are commercially interesting for economic and/or environmental reasons. Typical cogeneration processes to produce these chemicals will be described in this review, and their performance will be characterized simultaneously. As fuel cells are the key element in the processes and the electrolytes employed in the fuel cells are well reported in literature, cogeneration processes discussed in the following sections will be classified by their electrolytes and then further classified by the involved reactions. Such reactions are classified by their reactants and products. In each section, inorganic products will be discussed firstly, followed by organic products. The performance of these cogeneration processes will be characterized by the open circuit voltage, current efficiency, selectivity with respect to the chemicals obtained, current and power density, etc.

2.2 ELECTRICITY AND CHEMICAL COGENERATION PROCESSES

2.2.1 INTRODUCTION TO COGENERATION

One of the key reactions for high-value-added chemical synthesis is selective oxidation. However, such oxidation reactions induce two aspects of difficulties [6]. On the one hand, it is difficult to design suitable reactors and catalysts for a selective oxidation, as oxidation is easy to complete. Thus, selective oxidation requires favorable thermodynamic and kinetic conditions. Additionally, suitable catalysts should be carefully selected and tested for the process. On the other hand, the free energy change of a selective oxidation is much lower than that of a complete reaction. A particular example is that the oxidation of methane to ethane can provide only 150 kJ mol^{-1} (methane) free energy change, while 581 kJ mol^{-1} can be given by completely oxidizing methane. The lower energy provided by a selective oxidation than the corresponding complete oxidation requires precise heat management and high energy recovery efficiency. The high selectivity and high energy recovery efficiency of fuel cells as introduced previously as well as the flexible scales of fuel cells [7] make such technique very promising for fine chemical synthesis. Given such synthesis process generates electricity power simultaneously with chemical production, it is often referred to as electricity-chemical cogeneration. In order to operate the cogeneration devices to suitably generate electric power and value-added chemicals simultaneously, special adjustment to conventional fuel cells must be applied. For instance, anti-peroxide membranes must be fabricated for hydrogen peroxide cogeneration system. How to satisfy such special requirements has become a hotspot in fuel cell research.

Many efforts have been made to promote the development of fuel cell-type electricity-chemical cogeneration based on various types of fuel cells. The impetuses to develop cogeneration processes can be concluded into the following three aspects:

- to simplify the production of a chemical
- to achieve higher product yields
- to develop an environmental benign process.

$$\text{Oxygen Reduction:} \langle \text{I} \rangle n\text{O}_2 + 4ne^- \rightarrow 2n\text{O}^{2-} \langle /\text{I} \rangle \tag{2.1}$$

$$\text{Anode Oxidation:} \text{A} + 2n\text{O}^{2-} \rightarrow m\text{B} + 4ne^- \tag{2.2}$$

The reaction mechanism of the electricity and chemical cogeneration process is shown in Equations (2.1) and (2.2), which is similar to that involved in heterogeneous selective oxidation [8]. The similarity has been carefully studied by Garagounis et al., who reported that the function of oxygen reduction reaction (ORR) occurring in a solid fuel cell reactors is similar to that of oxygen in heterogeneous catalytic oxidation [4]. Thus, existing heterogeneous catalysts can provide some guidance for catalysts designed for cogeneration processes. Despite the similarity in reaction mechanism, cogeneration technique takes some significant advantages in energy conservation, economy, and environment, making it attract much attentions currently. The cogeneration technique can recycle large portions of the released chemical energy from the reaction by its fuel cell-type reactors, rather than wasting the energy in the form of low-grade thermal energy through heterogeneous catalytic oxidation process. And high selectivity can be achieved by separating the oxidation and reduction reactions in anode and cathode chambers, which are partitioned by selective permeable electrolyte membranes. Many electrolytes used in cogeneration processes are selectively conductive for ions that only targeted ions can pass through and contact with the reactants/generated ions on the other electrode and hence can avoid unexpected reactions. Additionally, since oxidants and fuels are supplied separately into the cathode side and anode side, the two reactants have no competition for catalytic active sites, which may lead to a higher reaction rate. The separated oxidants and fuels can largely reduce the risk of explosion as well. Furthermore, previous study reported that productivity can be controlled by external loads and in-feed flux of fuels and oxygen [9]; thus, cogeneration processes are rather adjustable.

2.2.2 CHEMICAL COGENERATION IN AQUEOUS ELECTROLYTE FUEL CELLS

Aqueous fuel cells utilizing alkaline, acid, or neutral aqueous electrolyte without using iron conductive membranes have been proposed for various applications, e.g., NO_x reduction [10,11], HCl production [12], partial oxidation of glycols [13–15], and partial oxidation of light alkanes [16]. One of the obvious advantages of using this kind of fuel cells for chemical cogeneration is the elimination of the complicated issues related to membrane conditioning [17,18].

The first attempt to utilize fuel cells for chemical cogeneration was made in reduction of nitric oxide. In this process, a mixture of N_2O, N_2, NH_2OH, and NH_3 was obtained through reactions (2.3), (2.4), (2.5), and (2.6) by introducing H_2 and NO to the anode and cathode compartments [11].

$$\text{Anode: } H_2 \rightarrow 2H^+ + 2e^- \quad E^0 = 0 V \tag{2.3}$$

$$\text{Cathode: } 2NO + 2H^+ + 2e^- \rightarrow N_2O + H_2O \quad E^0 = 1.59 V \tag{2.4}$$

$$2NO + 4H^+ + 4e^- \rightarrow N_2 + 2H_2O \quad E^0 = 1.68 V \tag{2.5}$$

$$2NO + 6H^+ + 6e^- \rightarrow 2NH_2OH \quad E^0 = 0.50 V \tag{2.6}$$

$$2NO + 10H^+ + 10e^- \rightarrow 2NH_3 + 2H_2O \quad E^0 = 0.84 V \tag{2.7}$$

NO conversion efficiency achieved was over 99.5% with 0.12 V cathode potential and 24 mA cm^{-2} current density, when platinum black cathode and sulfuric acid electrolyte were employed in the system. The high conversion rate of nitric oxide enabled the utilization of the chemical cogeneration process as a process for NO_x emission control. Additionally, the fraction of NH_3 in products ranges from 1.7% to as high as 86.5% accompanied by NH_2OH fractions between 0% and 2.6% by adjusting the cathode material, electrolyte, fuel feed rate, and external resistance. Subsequent studies successfully employed this mechanism for hydroxylamine production, which is mainly used in the synthesis of caprolactam by introducing proton-exchange membranes in the system. This system is classified in the type of polymer electrolyte fuel cells (PEFCs) and will be further discussed in the following section.

Another application of membrane-less fuel cells in cogeneration is in producing HCl by H_2/Cl_2 fuel cells. Though less popular than H_2/O_2 fuel cells, the literature on H_2/Cl_2 fuel cells was published as early as the 1920s [19,20], and catalysts, e.g., Pt [21,22], Pt–Ir alloy [23], RuO_2 [22], and $Pb_2Ru_2O_7$ [21], have been characterized for the purpose of Cl_2 reduction. As the reduction of Cl_2 is highly reversible compared with O_2 reduction, H_2/Cl_2 fuel cell can also be operated in the mode of Cl_2 and H_2 regeneration, which combined with its high power and energy density makes H_2/Cl_2 fuel cells an attractive candidate for space power application [21]. Thomassen and co-workers reported a fuel cell dedicated to HCl production, which consists of an electrode separator made by polyetheretherketone and current collectors made of nonporous graphite with HCl aqueous solution as the electrolyte [12]. The performance of this system was strongly dependent on the electrolyte concentration that the optimum performance was obtained using 3 M HCl at 50°C. Long-term performance of both Pt and Rh catalysts were also evaluated in this fuel cell. During the measurement, Rh catalyst saw a better long-term stability than Pt, while Pt provided a higher initial activity.

Partial oxidation of alcohols received much attention recently. The mechanisms of oxidation of ethanol with Pt catalysts and plurimetallic catalysts were studied by Vigier et al. [24]. In the case of ethanol partial oxidation to acetic acid and acetaldehyde at platinum electrode in acid electrolyte, a reaction mechanism with acetaldehyde intermediate product and acetic acid final product was proposed. The formation of acetic acid from ethanol is shown as Equations (2.8), (2.9), and (2.10).

$$Pt + H_2O \rightarrow Pt - OH + H^+ + e^- \tag{2.8}$$

$$Pt + CHO - CH_3 \rightarrow Pt - CO - CH_3 + H^+ + e^- \tag{2.9}$$

$$Pt - CO - CH_3 + Pt - OH \rightarrow 2Pt + CH_3 - COOH \tag{2.10}$$

However, the products of partial oxidation of light alcohols, such as methanol and ethanol, are mostly non-value added. In the case of value-added chemical cogeneration, polyols are interesting alternatives. Partial oxidation of glycerol in an alkaline fuel cell system was studied by Simões and co-workers [13]. The advantage of using glycerol as a fuel is that glycerol is becoming cheaper, as the increasing demand of methyl esters leads to an increase of by-produced glycerol production [25]. Non-platinum-based catalysts (Pd_xAu_{1-x}/C and $Pd_{0.5}Ni_{0.5}/C$) showed activity for partial oxidation of glycerol. Hydroxypyruvate ion, tartronate ion, and mesoxalate ion were detected in the products, revealing the possibility to apply non-platinum-based catalysts in alkaline medium to generate carboxylates. Recently, more non-platinum-based catalysts, Pd_1Sn_x and Pd_xBi, were reported by Zalineeva and co-workers [14,15]. Pd_1Sn_x showed high activity and selectivity for the production of C3 carboxylate compounds, and the existence of Sn inhibits the generation of CO_3^{2-} and CO_2; while the selectivity of Pd_xBi catalyst is adjustable by cell voltages that the catalyst is highly selective for aldehyde and ketone at low electrode potentials and at high electrode potentials, the catalyst can break the C–C bond and form CO_2 and carboxylates.

$$O_2 + 2H^+ + 2e^- (+catalysis) \rightarrow O^* (catalysis) + H_2O \tag{2.11}$$

$$O^* + R - H \rightarrow R\_OH(+catalysis) \tag{2.12}$$

2.2.3 CHEMICAL COGENERATION IN AQUEOUS PHOSPHOROUS ACID FUEL CELLS

The application of PAFCs in chemical cogeneration has been more or less forgotten recently. The latest work on the application of PAFCs cogeneration system was published a decade ago [16]. One of the importance to study PAFC-based cogeneration system appears to be the possibility to achieve selective oxidation at low temperatures.

A room temperature (298 K) system for partial oxidation of light alkanes in PAFCs was reported by Yamanaka and co-workers [16]. Partial oxidation of light alkanes into the corresponding oxygenates was achieved by a H_2/O_2 fuel cell system consisting of [light alkane, O_2|cathode\H_3PO_4/silica wool|anode|H_2, H_2O (g)]. In this system, O_2 and light alkane mixture was fed into the cathode chamber, while H_2 was fed into the anode chamber. As shown in Equation (2.11), O_2 is reduced and generates active oxy species (O*), while subsequently the generated active oxygen species (O*) oxidize hydrocarbons to corresponding oxygenates as shown in Equation (2.12). The foregoing reactions are rather sluggish under normal conditions. In order to promote the generation of active oxygen species to oxidize hydrocarbon, suitable electrocatalysis of the cathode is essential. Partial oxidation of propane into acetone,

acetone acid, and CO_2 was studied with various cathodes. A selectivity for acetone of 48% was obtained with a (Pd-black+VO(acac$_2$)/VGCF) cathode. However, oxidation of ethane on this cathode only obtained CO_2 and AcOH rather than other expected products, such as C_2H_5OH, CH_3CHO, CH_3OH, and HCHO, while in the oxidation of CH_4 on the same condition, a small amount of CO_2 is the only detected product.

2.2.4 Chemical Cogeneration in Polymer Electrolyte Fuel Cells

PEFCs employ polymer electrolyte membrane as the proton conductor. This type of fuel cells received much attention and funding in the recent couple of decades for their noteworthy features, such as low operating temperature, high power density, and easy scale-up [26]. Despite its application in power generation, efforts were also made to employ this technology to chemical production, such as hydroxylamine production [27–29], cyclohexylamine and aniline [30], hydrogenation [31,32], hydrogen peroxide production [33–35], and partial oxidation of hydrocarbon [36,37].

Industrial production of hydroxylamine is carried out by reduction of higher valent state nitrogen, including catalytic hydrogenation of nitric oxide, catalytic hydrogenation of nitrates, and the Raschig process [27]. Such techniques saw some critical disadvantages: the low selectivity and the usage of toxic materials. In order to overcome these shortcomings, efforts were made to explore the possibility of producing hydroxylamine using fuel cells. A nitric oxide-hydrogen fuel cell equipped with Pt electrodes, H_2SO_4 electrolyte, and proton-exchange membranes was reported [28]. The fuel cell obtains maximum current efficiencies of over 80%. In this system, high current efficiency and high selectivity for hydroxylamine were achieved by employing proton-exchange membranes, which were used to separate anode and cathode liquid compartments and only allow proton to pass through. Otherwise, the generated hydroxylamine would be reduced to ammonia after reaching the anode, leading to a decrease in current efficiency. The yield of hydroxylamine was sensitive to NO flow rate that low NO flow rate led to higher current efficiency. This phenomenon possibly can be explained by a consecutive reaction between generated hydroxylamine and bulk [Equation (2.13)] or the reaction between NO and HNO$^+$ [Equation (2.14) and (2.15)], both of which favor N_2O formation. These assumptions can also explain the higher current efficiency achieved by NO reduction accompanied by hydrogen. The generated hydroxylamine was neutralized by the acidic electrolyte and formed hydroxylammonium sulfate, which is more stable than hydroxylamine. Iron(II) phthalocyanine was studied as the catalyst for this process, which is considered to have the possibility to offer higher selectivity for hydroxylamine formation [29].

$$2NH_2OH + 4NO \rightarrow 3N_2O \uparrow + 3H_2O \qquad (2.13)$$

$$HNO^+_{ads} + NO + e^- \rightarrow HN_2O_{2(ads)} \qquad (2.14)$$

$$HN_2O_{2(ads)} + H^+ + e^- \rightarrow N_2O \uparrow + H_2O \qquad (2.15)$$

Cyclohexylamine and aniline were produced by selective reduction of nitrobenzene dissolved in ethanol in a PEMFC reactor [30]. The PEMFC consisted of a membrane electrode assembly (MEA), which was fabricated from Nafion 117 membranes incorporated with the anode and cathode and prepared using 20% Pt/C catalyst. H_2 was supplied into the anode chamber and oxidized to provide proton [Equation (2.16)]; cyclohexylamine and aniline were obtained in the cathode chamber through reactions (2.17) and (2.18). The by-product nitrocyclohexane might be formed by reaction (2.19). Cyclohexylamine and aniline were formed simultaneously only with external load. A selectivity of 57.3% to cyclohexylamine and a selectivity of 28.2% to aniline were reached with a nitrobenzene conversion efficiency of 8.2%. Both the selectivity of cyclohexylamine and the conversion efficiency of nitrobenzene decreased with the increase of hydrogen flow rate. This can be explained by a drying out effect of the membrane on the hydrogen side. The conversion of nitrobenzene and the selectivity for aniline increased by increasing Pt loading. The fuel cell performance was improved with the rise of temperature till 80°C, at which temperature the solvent starts to volatilize. The maximum power density achieved was $1.5\,mW\,cm^{-2}$ with a current density of $15\,mA\,cm^{-2}$.

$$\text{Anode: } 3H_2 \rightarrow 6H^+ + 6e^- \tag{2.16}$$

$$\text{Cathode: } C_6H_2NO_2 + 6H^+ + 6e^- \rightarrow C_6H_5NH_2 + 2H_2O \tag{2.17}$$

$$C_6H_5NH_2 + 6H^+ + 6e^- \rightarrow C_6H_{11}NH_2 \tag{2.18}$$

$$C_6H_5NO_2 + 6H^+ + 6e^- \rightarrow C_6H_{11}NO_2 \tag{2.19}$$

Later, hydrogenation in PEFCs was studied by the same group [31]. In this process, unsaturated organic compounds, e.g., allyl alcohol and maleic acid, were used as oxidants, while hydrogen was used as a fuel. The concept of electrogenerative hydrogenations was first studied by Langer et al. [32]. In order to study the mechanism of electrochemical hydrogenation, Yuan and co-workers studied the process with several techniques including cyclic voltammetry and high-performance liquid chromatography (HPLC). A two-step mechanism for cathode reaction was presented. First, hydronium ions are adsorbed onto Pt catalyst and obtain an electron [Equation (2.20)], and subsequently unsaturated organic compounds are absorbed onto Pt catalyst [Equation (2.21)] and hydrogenation occurs as reaction (2.22). This mechanism meets the results of their experiments well that only the hydrogenated products formed according to HPLC analysis.

Cathode:

$$Pt + H_3O^+ + e^- \rightarrow Pt - H + H_2O \tag{2.20}$$

$$nPt + R_1 - HC = CH - R_2 \rightarrow Pt_n - (R_1 - HC = CH - R_2)_{ad} \tag{2.21}$$

$$2Pt - H + Pt_n - (R_1 - HC = CH - R_2)_{ad} \rightarrow (n+2)Pt + R_1 - CH_2 - CH_2 - R_2 \tag{2.22}$$

where $R_2 = -CH_2-OH$, or $-COOH$.

Another important chemical studied to be produced by cogeneration process is hydrogen peroxide. Hydrogen peroxide is widely used as a kind of bleach, cleaning agent, or disinfectant for domestic, medical, and industrial use due to its strong oxidizing property. One of the most important applications of hydrogen peroxide is pulp and paper bleaching. In Europe, 41% of consumption of hydrogen peroxide is used for paper bleaching [38]. Additionally, hydrogen peroxide also finds a number of applications in the chemical industry for the synthesis of many organic and inorganic products [39,40], which make up for 43% of the amount of hydrogen peroxide usage in Europe [38]. Other applications of hydrogen peroxide include the usage in mining, metal processing, and rocketry. Hydrogen peroxide is considered an ideal "green oxidant" due to its high atom efficiency (47%) and the only theoretical coproduct, water [41]. Despite its diverse applications, the manufacture of hydrogen peroxide is almost exclusively dominated by Riedl-Pfleiderer or anthraquinone process, which was developed in 1936 and then patented in 1939 [38]. The huge centralized hydrogen plants required by such process have to be located at different sites, which are usually far from consumers. The consequent storage, delivery, and handling demands of this hazardous oxidant are dangerous for the local community and individuals.

Agladze and co-workers reported an alternative method for the production of hydrogen peroxide using DMFCs [33]. The possibility of employment of sea water, which was simulated by sodium chloride of various concentrations in their experiments, as a catholyte was studied. CH_3OH with KOH solution was supplied to the anode as a fuel, while air was used as an oxidant. Anolyte and catholyte were circulated during the operation. 1.1 V OCV was achieved with sea water as the catholyte and $5\,M\,KOH + 4\,M\,CH_3OH$ solution as the anolyte. However, the current efficiency to H_2O_2 decreased from 100% to 65% in the 3-hours operation, which can be explained by that the pores of cathode were blocked by carbonate precipitation generated by carbon dioxide absorption in the alkaline electrolyte. The improvement in the performance of electricity generation with the rise of NaCl concentration was observed. The increase in KOH concentration also positively influences the catalysts of methanol oxidation that the highest values of voltage and power generated by the fuel cell were achieved at 7 M KOH concentration. These improvements in performance are due to the increase in conductivity of either the catholyte or anolyte. The increase of methanol concentration from 0.5 to 0.1 M contributed to the rise of generated currency density from 7.5 to $10.8\,mA\,cm^{-2}$, while further increase of methanol concentration failed to achieve a further improvement in currency generation. The potential to improve the performance of the cogeneration process through rising methanol concentration is limited by the catalytically active sites of the anode. Additionally, the employment of methanol with higher concentrations may increase the possibility of CH_3OH penetration into the catholyte and hence consume the generated H_2O_2. The generation of hydrogen peroxide is negatively affected by the increase in temperature, as it accelerates the decomposition of hydrogen peroxide generated.

Sombatmankhong et al. developed another DMFC process to cogenerate hydrogen peroxide with electricity [34]. A MEA sandwiched between two gas diffusion layers (GDLs) was used in the system. Nafion 117 was used as a polymer electrolyte membrane. Methanol was supplied to the anode chamber as a fuel while oxygen was supplied to the cathode chamber as an oxidant. The half-cell electrochemical reactions are as follows:

$$\text{Anode: } CH_3OH + H_2O \rightarrow CO_2 + 6H^+ + 6e^- \qquad (2.23)$$

$$\text{Cathode: } O_2 + 2H^+ + 2e^- \rightarrow H_2O_2 \qquad (2.24)$$

In order to detect the generated peroxide, dichlorofluorescein (DCFH) was employed as a fluorescent probe, which can trace the presence of as low as 25 pmol peroxide [42]. It is notable that DCFH solution must be kept on ice and in dark place until use and discarded each day after use. The concentration of hydrogen peroxide can be determined by the fluorescence intensity. The correlation between average intensity and hydrogen peroxide concentration fitted by Cathcart and co-workers is given as

$$[H_2O_2]_{avg} = (6.61 \times 10^{-15}) \exp^{0.1625F}, \qquad (2.25)$$

where $[H_2O_2]_{avg}$ is the average concentration of hydrogen peroxide, while F is the average fluorescence intensity.

It is remarkable that the amount of hydrogen peroxide production and current efficiency decreased drastically with the increase of the cell potential. The maximum concentrations of hydrogen peroxide were 137.62, 23.67, and 5.44 μM with the corresponding current efficiencies of 85.00%, 64.78%, and 51.13% at the operating potentials of 300, 400, and 550 mV, respectively. A current density of 4.5–5.3 mA cm^{-2} and a power density of 0.8–1.2 mW cm^{-2} were obtained by the system.

Conventional hydrogen peroxide is produced through the cyclic reduction of oxygen by hydrogen using anthraquinone as a catalyst. This process, consisting of hydrogenation, oxidation, hydrogen peroxide extraction, and treatment of the working solution, usually requires huge plants at several sites. In contrast to the complicated production line, a novel small-size aluminum-air semi-fuel cell process has been developed with the potential for hydrogen peroxide cogeneration [35]. Different from normal fuel cells which use either liquid or gas fuel, the aluminum-air semi-fuel cell used aluminum as the reductant. Aluminum oxidization reaction at the cathode and two-electron ORR at the anode are maintained in this process. The electrode reactions and the total reaction are as the following:

$$\text{Anode: } 2Al + 8OH^- \rightarrow 2AlO_2^- + 4H_2O + 6e^- \quad E^0 = -2.35 \text{ V} \qquad (2.26)$$

$$\text{Cathode: } 3O_2 + 3H_2O + 6e^- \rightarrow 3HO_2^- + 3OH^- \quad E^0 = -0.065 \text{ V} \qquad (2.27)$$

$$\text{Total: } 2Al + 3O_2 + 5OH^- \rightarrow 2AlO_2^- + H_2O + 3HO_2^- \quad E^0 = -2.25 \text{ V} \qquad (2.28)$$

Figure 2.3 shows the scheme of the Al-air semi-fuel cell (SPC) for hydrogen-peroxide cogeneration. The power generated ranges from 0.058 to 8 mW cm^{-2} with external resistance from 10^4 to 10^1 Ω when using BP 2000 (supplied by ElectroCell

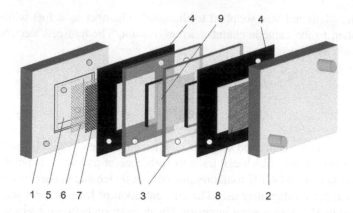

FIGURE 2.3 Principle scheme of the Al-air SPC for hydrogen peroxide cogeneration: (1 and 2) end plates, (3) gas-diffusion cathode, (4) nickel mesh, (5) gas chamber for air supply, (6) aluminum anode, (7) spacers for catholyte and anolyte supply, (8) ion-exchange membrane, and (9) rubber gaskets. (With permission from Agladze, G. et al., *J. Power Sources*, 218, 46, 2012.)

AB, Sweden) as cathode material, 1 M KOH electrolyte, and MA-40 (supplied by NPO "Schekino," Russia) as anion exchange membrane. The maximum current density achieved was $100\,\text{mA cm}^{-2}$. Hydrogen peroxide was generated at the rate of $58.3\,\text{mg h}^{-1}$ with a current efficiency of 85%. The performance of this process was influenced by the composition and concentration of electrolytes largely. The maximum power output rose twice with a 2.1 times higher current density was achieve by using 6 M KOH electrolyte. This can be explained by the improvement in the specific conductivity of the electrolytes. Poorer performance was observed at the use of 1 M NaOH, by which the maximum power density obtained was only $10.44\,\text{mW cm}^2$. This is caused by the less conductivity of the NaOH electrolyte. Additionally, the precipitation of carbonate due to absorption of CO_2 by alkaline electrolyte may cause pores of the gas diffusion layer to be blocked and consequently a reduction in the performance in power and hydrogen peroxide cogeneration. This is even more serious for cogeneration processes using NaOH electrolyte, due to the less solubility of sodium carbonate.

H_3PO_4-doped polymer electrolytes have been developed for partial oxidization of hydrocarbon. This kind of electrolytes can be operated at higher temperatures than conventional PEMFCs; thus, higher reaction kinetics and catalytic activities can be obtained, while the poisoning effect of the catalysts by CO is reduced [36]. H_3PO_4-doped polymer membranes become proton conductive at temperatures up to 200°C. The conductivity can be explained by the hopping mechanism that protons migrate along the anionic chain. Dehydrogenation of propane is a possible routine to cogenerate propylene and hydrogen with electricity. The reaction is expressed as follows:

$$C_3H_8 \rightarrow C_3H_6 + H_2 \tag{2.29}$$

However, in order to promote the conversion, hydrogen must be removed from the gaseous mixture as the equilibrium concentration of hydrogen is below 10% at temperatures below 250°C. The degree of oxidation strongly influences contents of the final products. Propane fuel cells with H_3PO_4-doped poly-benzimidazole polymer membranes generate oxygen-containing partial oxidation C_3 intermediate and then CO and CO_2 under light humidity at 250°C [37].

2.2.5 Chemical Cogeneration in Solid Oxide Fuel Cells

The cogeneration systems discussed above are all operated at low temperatures. However, the application of high-temperature cogeneration systems takes several advantages, including the possibility of internal fuel reforming [43], the wide range of fuel choices, and the minimization of polarization losses and of the poisoning effect of impurities such as H_2S [44]. Currently, SOFC-type reactors have received much more attention than any other type of fuel cell cogeneration systems. SOFCs have been proposed to generate a variety of chemicals with cogeneration of electricity. Nitric oxide [45], styrene [46], hydrogen cyanide [47], sulfur dioxide [48], carbon disulfide [49], hydrocyanic acid [50], syngas [51–56], C_2 hydrocarbons (ethane and ethylene [57]), hydrogen cyanide [50], and methanol [58] have been obtained by various reaction processes.

The feasibility to employ SOFCs in chemical cogeneration was first demonstrated in 1980. SOFCs were used in the oxidation of NH_3 to NO, combined with power generation [45]. Low-temperature fuel cell processes are infeasible for this purpose because N_2 is the only product of oxidation in that temperature range. The structure of the reactor is expressed as NH_3, NO, N_2, Pt/ZrO_2 (8% $Y2O_3$)/Pt, air. The cogeneration system consisted of an 8% yttria-stabilized zirconia (YSZ) tube. The conversion of NH_3 to NO is a highly exothermic reaction that the Gibbs free energy ΔG of this reaction is −64.5 kcal mol^{-1} of NH_3 at 1000 K. The cogeneration process captures this energy in the form of electricity rather than thermal energy, with the cogeneration of NO [Equation (2.30)], which is an essential ingredient for NHO_3 manufacture. However, the generated NO may react with NH_3 further, forming N_2 and reducing NO selectivity [Equation (2.31)]. The performance of the cogeneration process, i.e., selectivity of NO and power density, was highly dependent on NH_3 flow rate and operating temperature. High selectivity was achieved at the temperature between 900 and 1200 K. The highest selectivity to NO was 97% with a conversion rate of 24% and a power output of only 7 µW cm^{-3} at low NH_3 flow rate. The highest power output of the order of 10^{-3} W cm^{-3} solid electrolyte was achieved at high NH_3 flow rate. However, under this condition, the selectivity of NO is less than 5%.

Anode:

$$2NH_3 + 5O^{2-} \rightarrow 2NO + 3H_2O + 10e^- \qquad (2.30)$$

$$2NH_3 + 3NO \xrightarrow{k} 5/2N_2 + 3H_2 + 3H_2O \qquad (2.31)$$

Another early process using SOFCs is the cogeneration of hydrogen cyanide from a mixture of methane and ammonia [47]. The conversion of H_2S to SO_2 using SOFCs was achieved later [48].

Many efforts were made to employ SOFCs as electrochemical partial oxidation (EPOX) reactors. In this process, hydrocarbon fuel gases, e.g., CH_4, are partially oxidized into CO and H_2, i.e., syngas (2.32). Compared to conventional steam reforming, partial oxidation of methane can provide a better H_2/CO ratio. Additional, partial oxidation of methane, but not steam reforming, is an exothermic reaction with higher reaction rate. The SOFCs used by Galvita and co-workers consisted of a tube made of an YSZ electrolyte and Pt and Ni electrodes deposited on the outer and inner surface of the tube, respectively [51]. A CO selectivity of 95% and a CH_4 conversion efficiency over 90% were achieved at 850°C. The direct partial oxidation mechanism as shown in Equation (2.32) was supported by their research. Perovskite support with doped noble metals (Pt, Pd, Rh) was fabricated in order to improve catalyst performance in terms of activity and stability at high temperatures, as the catalytic performance of noble metals is highly dependent on the size of metal particles and the distribution [59,60]. $Rh-LaCoO_3/Al_2O_3$ and $Rh/LaMnO_3/Al_2O_3$ were employed in syngas cogeneration system [54]. Wiyaratn and co-workers employed $Au/LaSrMnO_3$ in syngas generation [61]. With the presence of Au, the CH_4 conversion efficiency was increased from 27.9% to 40.3% and the H_2 yield was increased from 20.5% to 30.6% compared to $LaSrMnO_3$ without doped Au, at the operating temperature of 1000°C and gas hourly space velocity of 7860 h^{-1}.

$$CH_4 + O^{2-} \rightarrow CO + 2H_2 + 2e^- \qquad (2.32)$$

A single-chamber SOFC (SC-SOFC) was developed for syngas generation [53]. The principle of SC-SOFCs is based on the different activity and selectivity of reactions occurring on the anode and cathode. Since the structure of this system is much more simplified than conventional dual-chamber SOFCs, a much more stable performance of SC-SOFCs can be expected. A samaria-doped ceria (SDC) electrolyte incorporated with a 10 wt % SDC-containing Ni anode and a $Sm_{0.5}Sr_{0.5}CoO_3$ cathode was developed for the SC-SOFC system, which was reported to have a much higher ionic conductivity than YSZ in an oxidizing atmosphere and to be more effective at lower temperatures than SOFCs with YSZ electrolyte [53]. The highest power density generated by the cell was ca. 400 mW cm^{-2} with a current density of around 900 mA cm^{-2} in a flowing mixture of ethane and air at 773 K, when employing a 0.15 mm SDC. Later, both the performances of electricity generation and syngas production were improved by employing bilayer electrolytes in the system and combustion-synthesized catalyst in the downstream [56]. The bilayer electrolyte was constructed by a layer of SDC incorporated combined with a layer of YSZ (Figure 2.4). A maximum power density of around 1500 mW cm^{-2} was achieved by a cell with 5 μm YSZ and 5 μm SDC bilayer electrolytes at a furnace temperature of 700°C operated in a flowing methane-oxygen gas mixture with a ratio of 2:1. After adopting a combustion-synthesized $GdNi/Al_2O_3$ catalyst in the downstream of the cell, a methane conversion of more than 95% and a H_2/CO ratio of around 2.0 were reached at a furnace temperature of 850°C. The combustion-synthesized catalyst applied in the downstream solved the problem of low performance in syngas generation by the bilayer electrolyte fuel cells. Several different cathode materials were characterized and compared for SC-SOFCs [62]. The conductivities of the

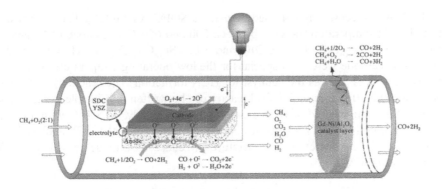

FIGURE 2.4 The configuration of the SC-SOFCs with a bilayer electrode. (With permission from Shao, Z. et al., *Angew. Chem. Int. Ed.*, 123, 1832, 2011.)

tested materials are in the following sequence: SSC>LSCF>LSM>BSCF (Table 2.2). When CH_4/O_2 was lower than 1, no degradation was observed in all of the tested cathode materials. However, when the CH_4/O_2 ratio = 1.5 was used, SSC and BSCF is degraded, which indicated that SSC and BSCF are not stable in high reducing atmosphere. At the furnace temperature ranging from 600 to 700°C, LSCF had the highest catalytic activities, followed by LSM, BSCF, and SSC. The catalytic activities of all these cathode materials increased with temperature and with CH_4/O_2 ratio, and hence a higher OCV was achieved. However, a higher OCV doesn't necessarily result in a better performance, because of the higher polarization resistance of the electrode at lower operating temperatures.

Electrochemical reactors based on fuel cell technology are particularly interesting for electrolysis of water to meet the increasing consumption of hydrogen. A large portion of the cost in conventional electrolysis of water is made up by the price of electricity, as the efficiency of the electrolysis process is relatively low. High-temperature fuel cell combined technology, such as SOFCs, has the potential to generate hydrogen with higher efficiency and hence reduce the usage of external electric power, enabling electrolysis of water to be cost competitive with reforming technologies [63]. Carbon fuel cell technology-assisted electrolyzers have been developed by Ewan and Adeniyi [64]. Carbon fuel is supplied to the cell anode, reducing the required electrolysis voltage by ca. 1 V.

TABLE 2.2

Materials Tested for Cathodes

Name	Composition	Supplier	d_{50} (µm)	S_{BET} (m² g⁻¹)
LSM	$La_{0.8}Sr_{0.2}MnO_{3-\delta}$	St Gobain	0.7	7.85
BSCF	$Ba_{0.5}Sr_{0.5}Co_{0.8}Fe_{0.2}O_{3-\delta}$	Marion Technologies	10	0.85
SSC	$Sm_{0.5}Sr_{0.5}CoO_{3-\delta}$	Fuel Cell Material	1	9.7
LSCF	$La_{0.6}Sr_{0.4}Co_{0.2}Fe_{0.8}O_{3-\delta}$	Fuel Cell Material	3	3.9

Source: With permission from Rembelski, D. et al., *Fuel Cells*, 12, 256, 2012.

HCN was successively obtained by using a SOFC system [50]. The system was tested at the temperature ranging between 500 and 650°C. An anode of Ni and an electrolyte of $Ce_{0.9}Gd_{0.1}O_{1.95}$ (CGO10) and a $La_{0.6}Sr_{0.4}Co_{0.2}Fe_{0.8}O_{3-\delta}$ (LSCF) CGO10 composite cathode were used special for the low operating temperature. Iron antimony oxide was used as the catalyst for HCN formation. The reactions in an oxygen ion-conducting electrochemical reactor can be written in principle as (2.33) and (2.34). The highest yield of HCN from methanol was 40% at 500°C with a corresponding selectivity of 47.5%. At this operating temperature, the OCV achieved was 0.92 V. Any further increase in temperature will cause a drastic decrease in the yield of HCN and OCV.

$$\text{Cathode: } 4e^- + O_2 \rightarrow 2O^{2-} \qquad (2.33)$$

$$\text{Anode: } CH_3OH + NH_3 + 2O^{2-} \rightarrow HCN + 3H_2O + 4e^- \qquad (2.34)$$

Cheng and co-workers reported a SOFC process running on 5% H_2S/95% CH_4 gaseous mixture [49]. CS_2 and sulfur were obtained in this process combined with simultaneous cogeneration of electricity. CS_2 was proposed to be formed by the reaction between methane and hydrogen sulfide [see Equation (2.35)]. In order to operate in the high H_2S concentration atmosphere, a strontium-doped lanthanum vanadate with a nominal composition of $La_{0.7}Sr_{0.3}VO_3$ (LSV) was used as the anode, which was proved to have good chemical resistance to H_2S. The maximum power density was 280 mW cm^{-2} at the operating temperature of 95°C. The conversion efficiency to CS_2 and the conversion efficiency to sulfur were 23.7% and 45.3%, respectively, with an output voltage of 0.7 V and a current density of 160 mA cm^{-2}. Despite generation of CS_2, a valuable chemical, the development of this SOFC system also enabled electricity supply to natural gas drilling sites for daily operation.

$$CH_4 + 2H_2S \rightarrow CS_2 + 4H_2 \qquad (2.35)$$

A SOFC process to produce C_2 hydrocarbons by oxidative coupling of methane (OCM) was reported [65]. The OCM to ethylene is given in (2.36). A LaSrMnO$_3$ cathode//YSZ//Au/LSM anode fuel cell was employed in this system. The effective area of anode is 1.9 cm^2. The schematic diagram of the SOFC is shown in Figure 2.5. When methane was fed into the anode chamber and air was fed into the cathode chamber, the electromotive force (e.m.f) and the closed circuit current were 0.86 V and 9.8 mA, respectively, at 1123 K. At this operating temperature, a methane conversion efficiency of ca. 4.4% and a C_2 products selectivity of ca. 16.3% were achieved with a significant CO generation. The low methane conversion efficiency and C_2 products selectivity reveal that better anodes are necessary for the employment of SOFCs in C_2 hydrocarbons production.

$$2CH_4 + O_2 \rightarrow C_2H_4 + 2H_2O \qquad (2.36)$$

Nagao and co-workers developed a process for direct oxidation of methane to methanol [58]. This system employed a SOFC with $Sn_{0.9}In_{0.1}P_2O_7$ electrolyte, which can be

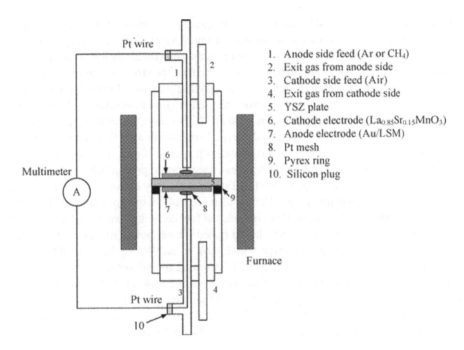

1. Anode side feed (Ar or CH$_4$)
2. Exit gas from anode side
3. Cathode side feed (Air)
4. Exit gas from cathode side
5. YSZ plate
6. Cathode electrode (La$_{0.85}$Sr$_{0.15}$MnO$_3$)
7. Anode electrode (Au/LSM)
8. Pt mesh
9. Pyrex ring
10. Silicon plug

Multimeter

A

Furnace

Pt wire

Pt wire

FIGURE 2.5 Schematic diagram of the SOFC for OCM. (With permission from Wiyaratn, W. et al., *J. Ind. Eng. Chem.*, 18, 1819, 2012.)

operated at lower temperatures (250°C or lower) than conventional SOFCs. 10 mol% In^{3+}-doped SnP$_2$O$_7$ showed high proton conductivities of above 10^{-1} S cm^{-1} between 100 and 350°C under water-free conditions. Methane–oxygen gaseous mixture with a ratio of 1:1 and hydrogen were supplied to the cathode and anode, respectively. Activated oxygen species were generated at the cathode with noble metal catalyst (2.38), and then the generated activated oxygen species directly oxidize methane into (2.39).

$$\text{Anode: } H_2 \rightarrow 2H^+ + 2e^- \tag{2.37}$$

$$\text{Cathode: } O_2 + 2H^+ + 2e^- \rightarrow O + H_2O \tag{2.38}$$

$$CH_4 + O \rightarrow CH_3OH \tag{2.39}$$

Recently, computational models received increasing attention in order to simulate EPOX process and to scale the process from the laboratory scale to practical application [66,67]. A model was built by Zhu et al. to simulate a 25 cm long anode-supported tubular cell with an inner diameter of 0.8 cm, a wall thickness of approximately 1175 μm, a YSZ electrolyte of 10 μm, and a porous LSM-YSZ cathode of 40 μm [66]. The initial 7.5 cm tube wall was composed by a barrier layer of 400 μm and Ni-YSZ of 725 μm. Inlet fuel and air velocities significantly influenced the EPOX performance. The maximum of syngas output was achieved at the fuel cell inlet velocity of 60 cm s^{-1} and the air velocity of 720 cm s^{-1}. Increasing velocity reduced the resident time, while decreasing velocity caused a portion of generated syngas consumed. Thus, either increasing or decreasing velocity decreased syngas output. At the optimized inlet velocities, the temperatures were generally lower and the temperature gradients were smoother, which can relieve the stress in the materials and hence alleviate degradation or damage. The original purpose of barrier layer was to capture generated steam within the porous anode to maintain sufficiently high steam-carbon and hence avoid carbon deposition on anode. However, since steam–carbon ratio increased along the channel, the use of barrier is relaxed at a particular length. A full barrier reduced overall EPOX performance, as it caused higher transport resistance and hence required larger gas-phase species concentration gradients between the fuel channel and the dense electrolyte. Fuel cell stacks were only one of the components in chemical cogeneration processes. The performance of balance-of-plant (BOP) components, i.e., the process taking place outside the fuel cell stack, was simulated by Lee and Strand [67]. Cathode gas recirculation raised the net power supply, as it reduced the power consumption of air blower, which took up ca. 95% of the total BOP power consumption. Anode gas recirculation also increased the net power output, since the available amount of fuel gases rose. Internal reforming took advantages on net power supply over external reforming due to the fact that the highly endothermic reforming process taking place inside the fuel cell decreases the stack temperature, reducing the demand for cooling air and hence reducing the power consumption of air blower.

2.2.6 CHEMICAL COGENERATION IN MOLTEN SALT FUEL CELLS

The partial oxidation of ethanol to acetaldehyde using molten salt fuel cells (MSFCs) was developed. This process is based on supported molten-salt catalysis that reactions are catalyzed in the liquid phase by solutions of transition metal ions or their coordination compounds. However, the catalyst utilization in conventional homogeneous catalysis is inefficient because of the limited gas–liquid interfacial area. Fuel cells using supported molten-salt catalysis can achieve large gas–liquid interfacial area by coating a molten-salt with the dissolved or dispersed transition metal complex catalyst onto the walls of a porous support (Figure 2.6). The gaseous phases diffuse through the left gas pore space and through the coated super-thin molten-salt phase and react within the catalyst layer. Several solvents were studied by Malhotra et al. [68], among which tetra-n-butylammonium trichlorostannate provided the best performance. A current density of 3.5 mA cm^{-2} with 0.1 V OCV and a product selectivity of 83% to acetaldehyde were achieved with tetra-n-butylammonium trichlorostannate as a melt for dissolving the catalyst at 90°C [68].

2.3 CONCLUSION

A review of current developments in fuel cells system applied in electricity and useful chemical cogeneration has been presented. Electro-cogeneration processes are a promising technology for fine chemical production. Compared to conventional fine chemical production through heterogeneous process, electro-cogeneration systems take advantages on energy conservation, economy, environment, etc. Since the first electro-cogeneration system was developed in the early 1980s, different fuel cell

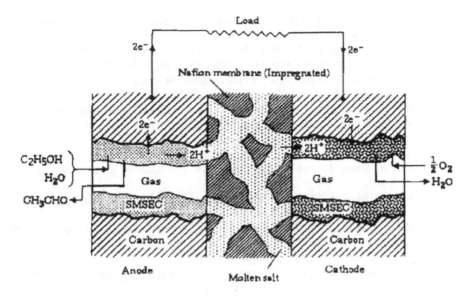

FIGURE 2.6   Schematic diagram of a supported molten-salt electrocatalytic fuel cell. (With permission from Malhotraa, S., Dam, R. 1996. Homogeneously catalyzed acetaldehyde and electricity from ethanol. *J. Electrochem. Soc.* 143:3058–3065.)

devices have been successfully employed in electro-cogeneration, including aqueous electrolyte fuel cells, PEFCs, PAFCs, SOFCs, MSFCs, etc. The large number of academic studies on the SOFC system appearing in the recent literature indicates that high-temperature fuel cell technology has the ability to convert fuels to various useful chemicals. The topic of low-temperature technologies, i.e., AEFCs, PEFCs, and PAFCs, is less studied recently, which is shown in the limited number of recent articles in this field. However, the earlier reported low-temperature electro-cogeneration systems have already proved the feasibility of this technology in chemical generation. Generally, electro-cogeneration technology is still in its primary stage. Further studies in such fields are desirable because many different electro-cogeneration processes that have been developed appear to be possible in practice, with adequate selectivity to desired chemicals and extra electric output, and cogeneration technology is possible to become an alternative means of chemical or electrolytic reactors.

REFERENCES

1. Boyce, M.P. 2006. *Gas Turbine Engineering Handbook*. Elsevier, Amsterdam.
2. Giddey, S. et al. 2012. A comprehensive review of direct carbon fuel cell technology. *Prog. Energy Combust. Sci.* 38:360–399.
3. Carrette, L., Friedrich, K., Stimming, U. 2000. Fuel cells: Principles, types, fuels, and applications. *ChemPhysChem* 1:162–93.
4. Garagounis, I. et al. 2011. Solid electrolytes: Applications in heterogeneous catalysis and chemical cogeneration. *Ind. Eng. Chem. Res.* 50:431–472.
5. Alcaide, F., Cabot, P.-L., Brillas, E. 2006. Fuel cells for chemicals and energy cogeneration. *J. Power Sources* 153:47–60.

6. Tagawa, T. et al. 1999. Fuel cell type reactor for chemicals-energy co-generation. *Chem. Eng. Sci.* 54:1553–1557.

7. Ellis, M.W., von Spakovsky, M.R., Nelson, D.J. 2001. Fuel cell systems: Efficient, flexible energy conversion for the 21st century. *Proc. IEEE* 89:1808–1818.

8. Guo, Z. et al. 2014. Recent advances in heterogeneous selective oxidation catalysis for sustainable chemistry. *Chem. Soc. Rev.* 43:3480–524.

9. Brillas, E., Alcaide, F., Cabot, P.L. 2002. A small-scale flow alkaline fuel cell for on-site production of hydrogen peroxide. *Electrochim. Acta* 48:331–340.

10. Otsuka, K. 1996. A hydrogen-nitric oxide cell for the synthesis of hydroxylamine. *J. Electrochem. Soc.* 143:3491–3492.

11. Langer, S., Pate, K. 1980. Electrogenerative reduction of nitric oxide. *Nature* 284:434–435.

12. Thomassen, M. et al. 2003. H_2 / Cl_2 fuel cell for co-generation of electricity and HCl. *J. Appl. Electrochem.* 33:9–13.

13. Simões, M., Baranton, S., Coutanceau, C. 2010. Electro-oxidation of glycerol at Pd based nano-catalysts for an application in alkaline fuel cells for chemicals and energy cogeneration. *Appl. Catal. B Environ.* 93:354–362.

14. Zalineeva, A. et al. 2015. Glycerol electrooxidation on self-supported Pd_1Sn_x nanoparticules. *Appl. Catal. B Environ.* 176–177:429–435.

15. Zalineeva, A. et al. 2014. Self-supported Pd_xBi catalysts for the electrooxidation of glycerol in alkaline media. *J. Am. Chem. Soc.* 136:3937–3945.

16. Yamanaka, I., Hasegawa, S., Otsuka, K. 2002. Partial oxidation of light alkanes by reductive activated oxygen over the (Pd-black + $VO(acac)_2$ / VGCF) cathode of $H_2 - O_2$ cell system at 298 K. *Appl. Catal. A Gen.* 226:305–315.

17. Mousavi Shaegh, S.A., Nguyen, N.T., Chan, S.H. 2011. A review on membraneless laminar flow-based fuel cells. *Int. J. Hydrogen Energy* 36:5675–5694.

18. Shukla, A.K., Raman, R.K., Scott, K. 2005. Advances in mixed-reactant fuel cells. *Fuel Cells* 5:436–447.

19. Foerster, F., Nobis, A., Stötzer, H. 1923. Über die wasserstoff-chlor-kette. *Zeitschrift für Elektrochemie und Angewandte Physikalische Chemie* 29:64–79.

20. Schmid, A. 1924. Die diffusionsgaselektrode. *Helv. Chim. Acta* 7:370–373.

21. Anderson, B., Taylor, E.J., Wilemski, G. 1994. A high performance hydrogen /chlorine space power applications fuel cell. *J. Power Sources* 47:321–328.

22. Yeo, R.S. et al. 1980. An electrochemically regenerative hydrogen-chlorine energy storage system. *J. Electrochem. Soc.* 10:393–404.

23. Shibl, S.M.A., Noel, M. 1993. Platinum-iridium bimetal catalyst-based porous carbon electrodes for H_2-Cl_2 fuel cells. *Int. J. Hydrogen Energy* 18:141–147.

24. Vigier, F. et al. 2006. Electrocatalysis for the direct alcohol fuel cell. *Top. Catal.* 40:111–121.

25. Clacens, J.M., Pouilloux, Y., Barrault, J. 2002. Selective etherification of glycerol to polyglycerols over impregnated basic MCM-41 type mesoporous catalysts. *Appl. Catal. A Gen.* 227:181–190.

26. Wang, Y. et al. 2011. A review of polymer electrolyte membrane fuel cells: Technology, applications, and needs on fundamental research. *Appl. Energy* 88:981–1007.

27. Ritz, J., Fuchs, H., Perryman, H.G. 2000. Hydroxylamine. In *Ullmann's Encyclopedia of Industrial Chemistry*, Wiley-VCH, Weinheim, 503–508.

28. Lewdorowicz, W. et al. 2006. Synthesis of hydroxylamine in the nitric oxide—hydrogen fuel cell. *J. New Mater. Electrochem. Syst.* 9:339–343.

29. Alvarez-Gallego, Y. et al. 2012. Development of gas diffusion electrodes for cogeneration of chemicals and electricity. *Electrochim. Acta* 82:415–426.

30. Yuan, X.-Z. et al. 2001. Cogeneration of cyclohexylamine and electrical power using PEM fuel cell reactor. *Electrochem. Commun.* 3:599–602.
31. Yuan, X.-Z. et al. 2005. Cogeneration of electricity and organic chemicals using a polymer electrolyte fuel cell. *Electrochim. Acta* 50:5172–5180.
32. Langer, S.H., Landi, H.P., 1964. The nature of electrogenerative hydrogenation. *J. Am. Chem. Soc.* 86:4694–4698.
33. Agladze, G. et al. 2010. DMFC with hydrogen peroxide cogeneration. *J. Electrochem. Soc.* 157:E140–E147.
34. Sombatmankhong, K., Yunus, K., Fisher, A.C. 2013. Electrocogeneration of hydrogen peroxide: Confocal and potentiostatic investigations of hydrogen peroxide formation in a direct methanol fuel cell. *J. Power Sources* 240:219–231.
35. Agladze, G. et al. 2012. A novel aluminium-air semi-fuel cell operating with hydrogen peroxide co-generation. *J. Power Sources* 218:46–51.
36. Li, Q., Hjuler, H.A., Bjerrum, N.J. 2001. Phosphoric acid doped polybenzimidazole membranes: Physiochemical characterization and fuel cell applications. *J. Appl. Electrochem.* 31:773–779.
37. Cheng, C.K. et al. 2005. Propane fuel cells using phosphoric-acid-doped polybenzimidazole membranes. *J. Phys. Chem. B* 109:13036–13042.
38. Campos-Martin, J.M., Blanco-Brieva, G., Fierro, J.L.G. 2006. Hydrogen peroxide synthesis: an outlook beyond the anthraquinone process. *Angew. Chem. Int. Ed.* 45:6962–6984.
39. Sato, K., Aoki, M., Noyori, R. 1998. A "green" route to adipic acid: Direct oxidation of cyclohexenes with 30% hydrogen peroxide. *Science* 281:1646–1647.
40. Sanderson, W.R. 2000. Cleaner industrial processes using hydrogen peroxide. *Pure Appl. Chem.* 72:1289–1304.
41. Noyori, R., Aoki, M., Sato, K. 2003. Green oxidation with aqueous hydrogen peroxide. *Chem. Commun.* 1977–1986.
42. Cathcart, R., Schwiers, E., Ames, B.N. 1983. Detection of picomole levels of hydroperoxides using a fluorescent dichlorofluorescein assay. *Anal. Biochem.* 134:111–116.
43. Achenbach, E. 1994. Three-dimensional and time-dependent simulation of a planar solid oxide fuel cell stack. *J. Power Sources* 49:333–348.
44. Haga, K. et al. 2008. Poisoning of SOFC anodes by various fuel impurities. *Solid State Ionics* 179:1427–1431.
45. Vayenas, C.G., Farr, R.D. 1980. Cogeneration of electric energy and nitric oxide. *Science* 208:593–594.
46. Michaels, J.N., Vayenas, C.G. 1984. Styrene production from ethylbenzene on platinum in a zirconia electrochemical reactor. *J. Electrochem. Soc.* 131:2544–2550.
47. Kiratzis, N., Stoukides, M. 1987. The synthesis of hydrogen cyanide in a solid electrolyte fuel cell. *J. Electrochem. Soc.* 134:1925–1929.
48. Yentekakis, I.V., Vayenas, C.G. 1989. Chemical cogeneration in solid electrolyte cells: The oxidation of H_2S to SO_2. *J. Electrochem. Soc.* 136:996–1002.
49. Cheng, Z., Zha, S., Aguilar, L. 2006. A solid oxide fuel cell running on H_2S/CH_4 fuel mixtures. *Electrochem. Solid State Lett.* 9:A31–A33.
50. Raj, A., Rudkin, R.A., Atkinson, A. 2010. Cogeneration of HCN in a solid oxide fuel cell. *J. Electrochem. Soc.* 157:B719–B725.
51. Galvita, V.V. et al. 1997. Electrocatalytic conversion of methane to syngas over Ni electrode in a solid oxide electrolyte cell. *Appl. Catal. A Gen.* 165:301–308.
52. Ishihara, T. et al. 1999. Partial oxidation of methane over fuel cell type reactor for simultaneous generation of synthesis gas and electric power. *Chem. Eng. Sci.* 54:1535–1540.
53. Hibino, T. 2000. A low-operating-temperature solid oxide fuel cell in hydrocarbon-air mixtures. *Science* 288:2031–2033.

54. Cimino, S. et al. 2006. Rh–La(Mn,Co)O$_3$ monolithic catalysts for the combustion of methane under fuel-rich conditions. *Catal. Today* 117:454–461.
55. Wei, H.J. et al. 2008. Lattice oxygen of La$_{1-x}$Sr$_x$MO$_3$ (M=Mn, Ni) and LaMnO$_{3-\alpha}$F$_\beta$ perovskite oxides for the partial oxidation of methane to synthesis gas. *Catal. Commun.* 9:2509–2514.
56. Shao, Z. et al. 2011. Electric power and synthesis gas co-generation from methane with zero waste gas emission. *Angew. Chem. Int. Ed.* 123:1832–1837.
57. Kelly, I., Middleton, H., Rudkin, R. 1988. Oxidation of methane in solid state electrochemical reactors. *Solid State Ionics* 28–30:1547–1552.
58. Nagao, M. et al. 2006. A proton-conducting In^{3+}-doped SnP$_2$O$_7$ electrolyte for intermediate-temperature fuel cells. *Electrochem. Solid State Lett.* 9:A105–A106.
59. Haruta, M. 1997. Size- and support-dependency in the catalysis of gold. *Catal. Commun.* 36:153–166.
60. Naknam, P. et al. 2007. Preferential catalytic oxidation of carbon monoxide in presence of hydrogen over bimetallic AuPt supported on zeolite catalysts. *J. Power Sources* 165:353–358.
61. Wiyaratn, W. 2011. Development of Au/La$_{1-x}$Sr$_x$MnO$_3$ nanocomposites for further application in a solid oxide fuel cell type reactor. *J. Ind. Eng. Chem.* 17:474–478.
62. Rembelski, D. et al. 2012. Characterization and comparison of different cathode materials for SC-SOFC: LSM, BSCF, SSC, and LSCF. *Fuel Cells* 12:256–264.
63. Manage, M.N. et al. 2011. A techno-economic appraisal of hydrogen generation and the case for solid oxide electrolyser cells. *Int. J. Hydrogen Energy* 36:5782–5796.
64. Ewan, B., Adeniyi, O. 2013. A demonstration of carbon-assisted water electrolysis. *Energies* 6:1657–1668.
65. Wiyaratn, W. et al. 2012. Au/La$_{1-x}$Sr$_x$MnO$_3$ nanocomposite for chemical-energy cogeneration in solid oxide fuel cell reactor. *J. Ind. Eng. Chem.* 18:1819–1823.
66. Zhu, H. et al. 2008. Modeling electrochemical partial oxidation of methane for cogeneration of electricity and syngas in solid-oxide fuel cells. *J. Power Sources* 183:143–150.
67. Lee, K.H., Strand, R.K. 2009. SOFC cogeneration system for building applications, part 1: Development of SOFC system-level model and the parametric study. *Renew. Energy* 34:2831–2838.
68. Malhotraa, S., Dam, R. 1996. Homogeneously catalyzed acetaldehyde and electricity from ethanol. *J. Electrochem. Soc.* 143:3058–3065.

3 Photoelectrochemically Driven Electrosynthesis

Hisashi Shimakoshi and Yoshio Hisaeda
Kyushu University

CONTENTS

3.1   Introduction to Photoelectrochemistry 67
3.2   Advantages of Light-Assisted Electrosynthesis in Mediator System 68
3.3   Light-Assisted Electrosynthesis by Natural Cobalamin Derivative 69
3.4   Light-Assisted Electrosynthesis by Vitamin B_{12} Derivatives 71
3.5   Light-Assisted Electrosynthesis by Simple Vitamin B_{12} Model Complex 74
3.6   Outer or Inner Photovoltaic Device Systems 76
3.7   Concluding Remarks 78
References 78

3.1   INTRODUCTION TO PHOTOELECTROCHEMISTRY

Electrosynthesis is currently expanding into various interdisciplinary fields because of its versatile application to molecular transformations as well as its sustainable and economic advantages [1,2]. Electroorganic synthesis is especially recognized as a green process in synthetic organic chemistry since the redox process of a substrate occurs due to an electric current without any chemical reagent and waste after the reaction. As for modern electrosynthetic methods, the electrode process combined with light irradiation has been developed in which the light irradiation is categorized into three methods, as shown in Figure 3.1. Among the light-assisted electrosyntheses, the light-assisted reaction of a substrate has significantly progressed in past decades and has a synergistic effect on the electrosynthesis. For example, electrochemically generated species are changed to another one by light irradiation without any additional electric current so that the overall energy consumption could be reduced during the reaction. Besides an economic reason, an excessive applied potential is undesirable for reaction in order to avoid a side reaction. Another advantage of the photoelectrochemically driven electrosynthesis is utilization of photo-to electro-energy conversion systems equipped with power supply using external photovoltaic devices or an inner dye-sensitized cell system, as shown in Figure 3.1. Photo-energy is efficiently converted to electro-energy, and then produces chemicals

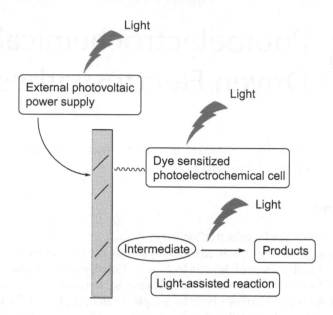

FIGURE 3.1 Photoelectrochemical process in organic synthesis.

by electrolysis. The advantages of photoirradiation in electrosynthesis are intro-duced in this chapter.

3.2   ADVANTAGES OF LIGHT-ASSISTED ELECTROSYNTHESIS IN MEDIATOR SYSTEM

Light-assisted electrosynthesis has been developed for indirect electrolysis using a mediator. Transition metal complexes can be excellent catalysts for electroorganic synthesis due to efficient electron transfer property with an electrode. Therefore, transition metal complexes are widely used as a mediator for indirect electroorganic synthesis [3]. One of the significant advantages of the indirect electrolysis associated with transition metal catalysis is being a convenient alternative to the usual methods to generate in situ low-valent species, which are not easily prepared and/or handled. For example, the electrochemically generated Co(I) species is known as a supernu-cleophile that forms an alkylated complex by reaction with electrophiles such as an organic halide (RX). The alkylated complex is a useful reagent for forming radical species as the cobalt–carbon bond is readily cleaved homolytically by electrolysis, thermolysis, and photolysis [3]. Therefore, the application of the alkylated complex to organic synthesis is quite interesting from the viewpoint of a radical-forming reagent in place of the conventional chemical reagent such as tin hydride. In the electrochem-ical system, the formation of radical species is facilitated during photoirradiation as the cobalt–carbon bond of the alkylated complex is cleaved by photolysis, as shown in Figure 3.2. Thus, electrolysis of an electrophilic reagent in the presence of a cobalt complex as a mediator will form a radical intermediate under a lower applied poten-tial with the support of light energy (*photochemical "shunt path"*). A variety of

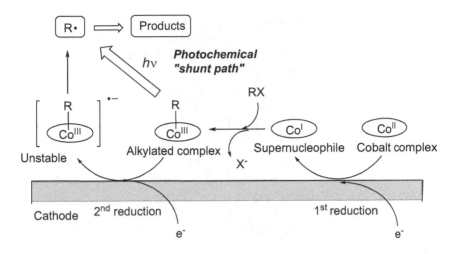

FIGURE 3.2 Photoelectrochemical organic synthesis by cobalt complex.

radical-mediated electroorganic syntheses mediated by cobalt complexes with light irradiation have been developed.

3.3 LIGHT-ASSISTED ELECTROSYNTHESIS BY NATURAL COBALAMIN DERIVATIVE

Vitamin B_{12} is a term found in health and nourishment studies, and has the chemical name cyanocobalamin [4]. However, this name is typically used for a cobalamin derivative. The vitamin B_{12} derivative denotes the cobalamin family in this chapter. Vitamin B_{12} consists of a cobalt atom coordinated to the tetrapyrrole ring system (corrin ring). One of the unique properties of vitamin B_{12} is possessing a stable cobalt-carbon bond in its biological function, and vitamin B_{12} is associated with radical-mediated enzymatic reactions. Due to this unique property, the application of vitamin B_{12} in catalytic chemistry was developed in electroorganic synthesis.

Natural cobalamin, hydroxycobalamin hydrochloride (vitamin B_{12a}) (see Chart 3.1) catalyzed the electrosynthesis under photoirradiation for the synthesis of 1,4-dioxo compounds (Equation 3.1) [5,6].

1,4-Dioxo compounds are valuable precursors for the synthesis of cyclopentanoids and furanoids. The reaction of the catalyst involves the formation and cleavage of a cobalt–carbon bond. The electrochemically formed Co(I) state of cobalamin at −0.95 V vs. SCE in dimethylformamide (DMF) reacts with carboxylic anhydrides to form the Co(III) acyl compound. Cleavage of the cobalt–carbon bond of the Co(III) acyl compound by visible light irradiation ($\lambda = 400$–550 nm) forms an acyl fragment, and it adds to the activated olefin to produce the desired 1,4-dioxo compound as shown in Figure 3.3. The mechanism of such a light-assisted nucleophilic acylation of activated olefins was kinetically investigated [7].

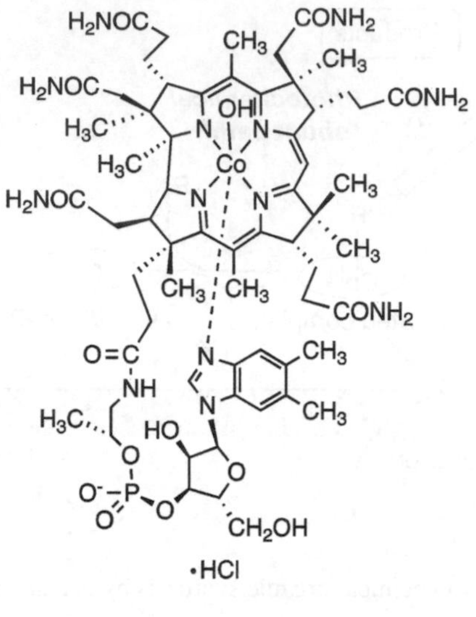

vitamin B$_{12a}$

CHART 3.1

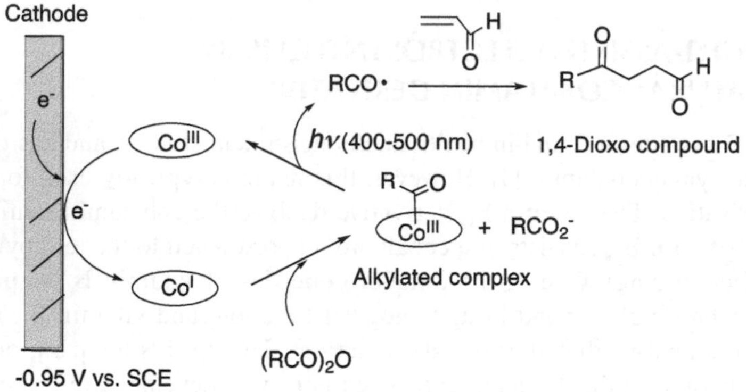

FIGURE 3.3　Electrochemical acylation of activated olefins mediated by vitamin B$_{12a}$ during photoirradiation.

The reductive radical cyclizations of bromo acetals and (bromomethyl)silyl ethers of terpenoid alcohols also occurred using vitamin B12a (Equations 3.2–3.4) [8]. Electrolysis at −1.2 V vs. SCE in DMF in the presence of catalytic amounts of vitamin B$_{12a}$ (3–5 mol%) during photoirradiation (halogen lamp, 250 W) produced cyclic products. The intramolecular cyclization was also applied to a sequential radical reaction. The electrolysis of 3-(2′-bromo-1′-ethoxy)cyclopentene with 1-cyanovinyl-acetate in the presence of vitamin B$_{12a}$ (0.75 mol%) at −1.1 V vs. SCE in DMF during photoirradiation (halogen lamp, 150 W) produced a bicyclic compound (Equation 3.5) [9]. This product is the precursor of methyl jasmonate or epituberolide.

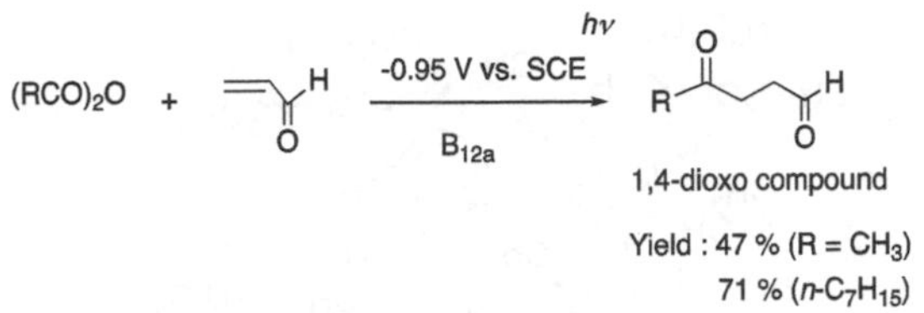

$$(RCO)_2O \ + \quad \xrightarrow[\substack{B_{12a}}]{\substack{h\nu \\ -0.95 \text{ V vs. SCE}}} \quad \quad \quad (1)$$

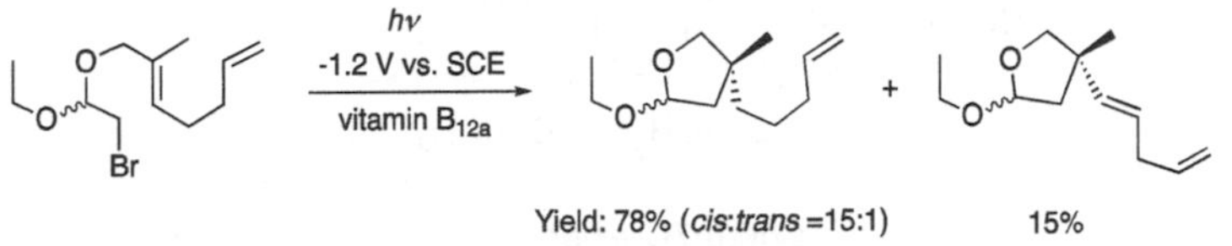

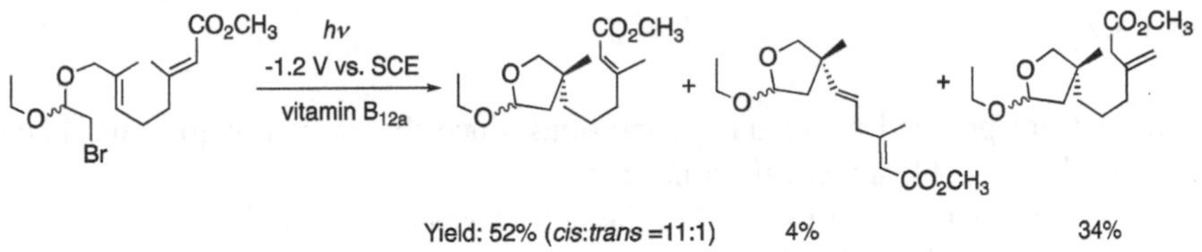

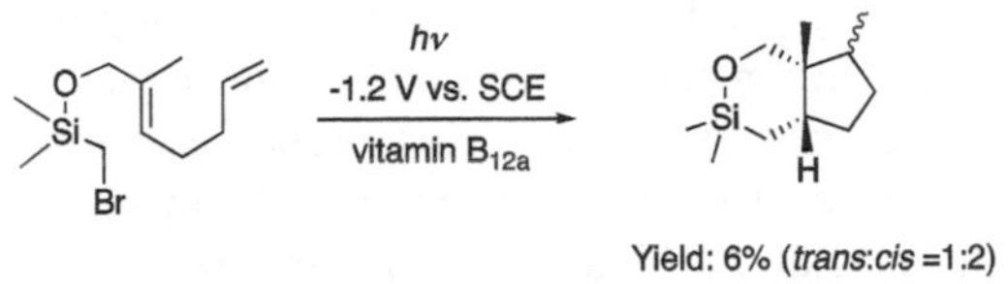

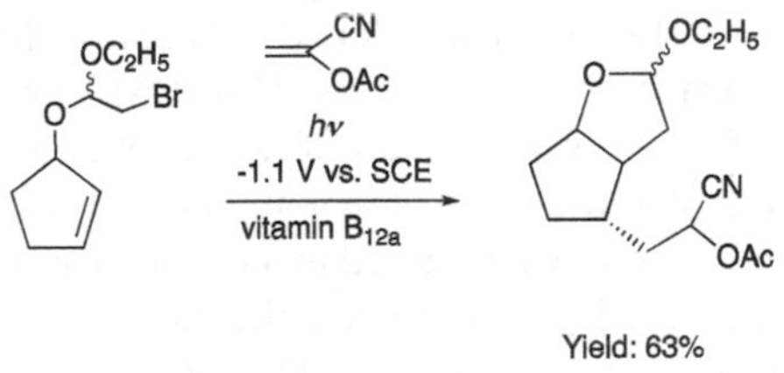

1,4-dioxo compound

Yield : 47 % (R = CH$_3$)

71 % (n-C$_7$H$_{15}$)

$$\xrightarrow[\substack{\text{vitamin } B_{12a}}]{\substack{h\nu \\ -1.2 \text{ V vs. SCE}}} \quad \quad + \quad \quad (2)$$

Yield: 78% (cis:trans =15:1)   15%

$$\xrightarrow[\substack{\text{vitamin } B_{12a}}]{\substack{h\nu \\ -1.2 \text{ V vs. SCE}}} \quad \quad + \quad \quad + \quad \quad (3)$$

Yield: 52% (cis:trans =11:1)   4%   34%

$$\xrightarrow[\substack{\text{vitamin } B_{12a}}]{\substack{h\nu \\ -1.2 \text{ V vs. SCE}}} \quad \quad (4)$$

Yield: 6% (trans:cis =1:2)

$$\xrightarrow[\substack{\text{vitamin } B_{12a}}]{\substack{h\nu \\ -1.1 \text{ V vs. SCE}}} \quad \quad (5)$$

Yield: 63%

3.4 LIGHT-ASSISTED ELECTROSYNTHESIS BY VITAMIN B$_{12}$ DERIVATIVES

Light-assisted electrosynthesis was also carried out using a vitamin B$_{12}$ derivative as the mediator. A heptamethyl cobrinate (Cob(II)7C$_1$ester) as shown in Chart 3.2 was synthesized from natural cobalamin and well reproduced the various properties of natural cobalamin such as redox behavior, reactivity toward a substrate, formation and cleavage of cobalt–carbon bond [3,10]. This vitamin B$_{12}$ derivative mediated the simple reduction of an organic bromide under light irradiation [11,12]. For example, the electrolysis of 2,2-bis(ethoxycarbonyl)-1-bromopropane was carried out at −1.0 V vs. SCE in DMF under visible light irradiation (tungsten lamp, 300 W) in the presence of a catalytic amount of heptamethyl cobrinate. The reduced product, 2,2-bis(ethoxycarbonyl)propane, was formed in 9–12% yields (Equation 3.6). The

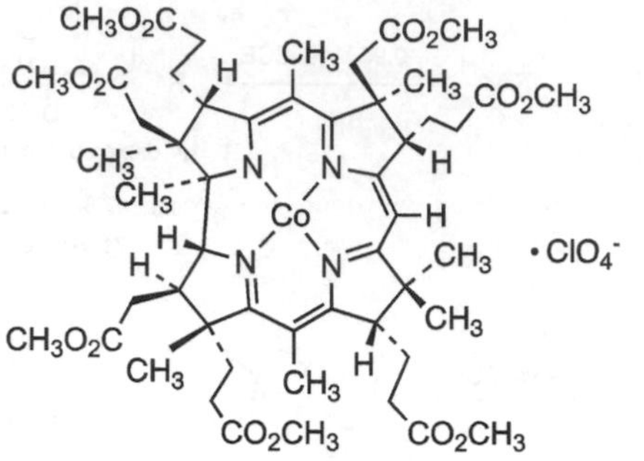

Cob(II)7C$_1$ester

CHART 3.2

reaction did not proceed under dark conditions since the corresponding alkylated cobalt complex is stable under this condition.

During the course of this study, the 1,2-migration of a functional group was achieved by light-assisted electrolysis with the vitamin B$_{12}$ derivative. The 1,2-migration of a functional group is a significantly important process in synthetic organic chemistry since the carbon-skeleton rearrangement creates various synthetic intermediates for fine chemical syntheses. This carbon-skeleton rearrangement reaction is mediated by adenosylcobalamin in methylmalonyl-CoA mutase, glutamate mutase, and methyleneglutarate mutase during an enzymatic process [13,14]. To mimic this enzymatic reaction, the heptamethyl cobrinate-mediated electrolysis under photoirradiation was effectively developed. The electrolysis of 2-acetyl-1-bromo-2-ethoxycarbonylpropane at −1.0 V vs. SCE in DMF under visible light irradiation in the presence of a catalytic amount of heptamethyl cobrinate affords the 10% yield of an acetyl-migrated product as the main product (Equation 3.7) [15].

The 1,2-migration of an acyl group was applied to a ring-expansion reaction. The electrolysis of alicyclic ketones (5-, 6-, 7-, and 8-membered rings) with a carboxylic ester and a bromomethyl group was carried out in DMF in the presence of a catalytic amount of heptamethyl cobyrinate at −1.0 V vs. SCE under photoirradiation (Equation 3.8) [16]. The ring expanded products were obtained in 5~11% yields.

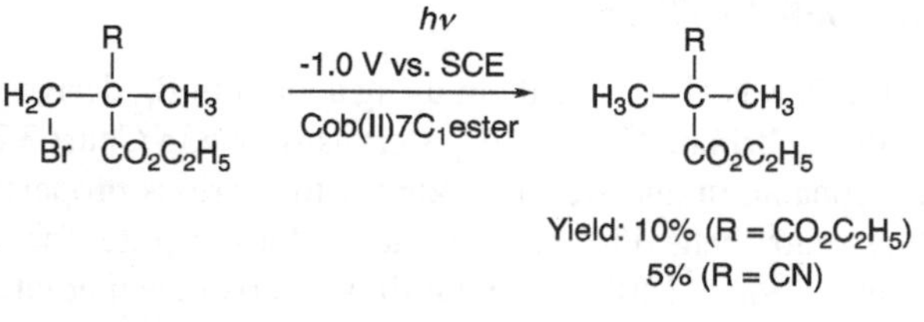

$$\text{Yield: 10\% (R = CO}_2\text{C}_2\text{H}_5)$$
$$\text{5\% (R = CN)}$$

$$\text{Yield: 10\%} \qquad\qquad 3\%$$

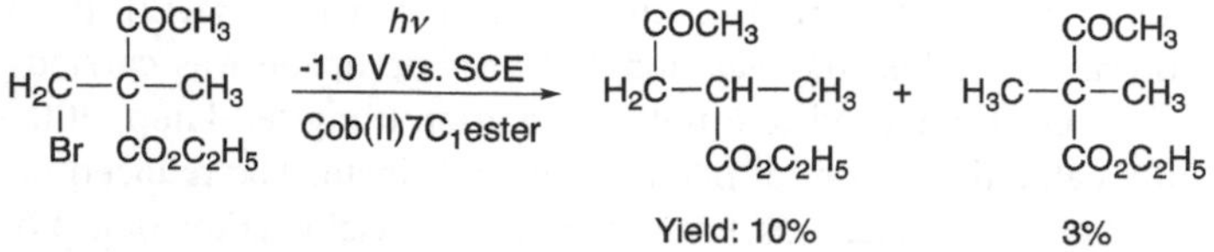

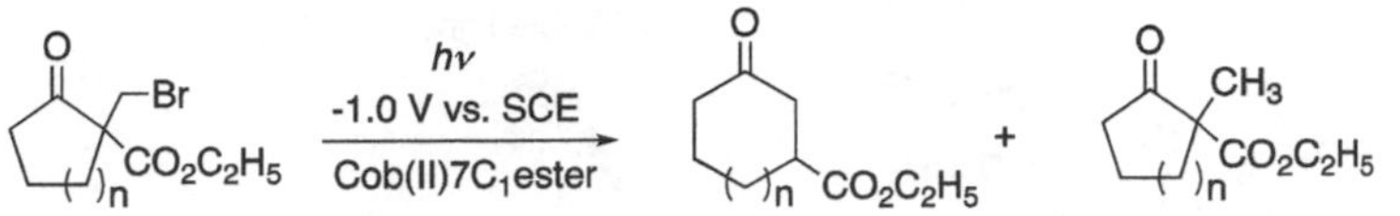

$$\begin{array}{c} hv \\ \text{-1.0 V vs. SCE} \\ \hline \text{Cob(II)7C}_1\text{ester} \end{array}$$

(8)

Yield

n = 1 (5-membered ring)   6%   1%
n = 2 (6-membered ring)   5%   3%
n = 3 (7-membered ring)   11%   1%
n = 4 (8-membered ring)   7%   1%

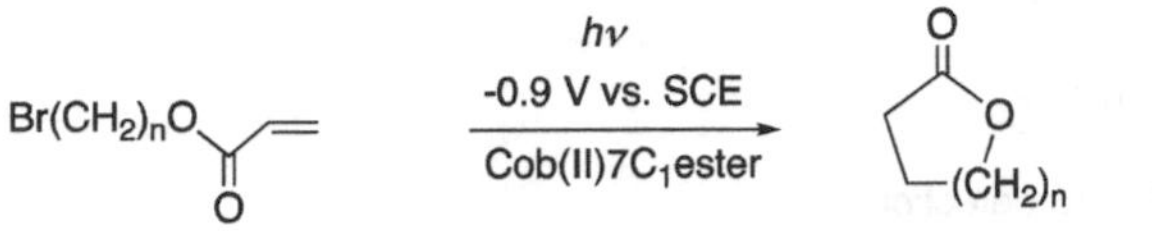

$$\begin{array}{c} hv \\ \text{-0.9 V vs. SCE} \\ \hline \text{Cob(II)7C}_1\text{ester} \end{array}$$

(9)

Yield: 42% (n = 12)
11% (n = 6)
5% (n = 2)

$$\begin{array}{c} hv \\ \text{-0.9 V vs. SCE} \\ \hline \text{TiO}_2/\text{B}_{12}(\text{COO}^-)_6 \end{array}$$

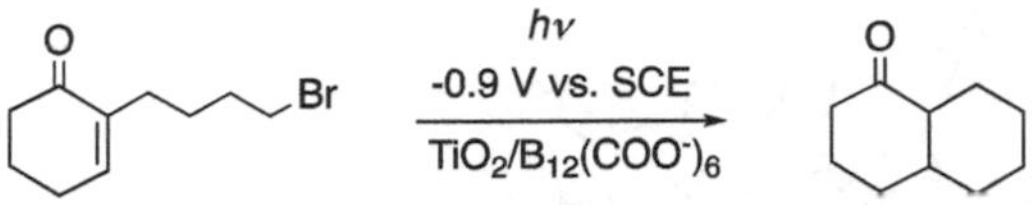

(10)

Yield: 65% (*trans:cis* =14:1)

Heptamethyl cobyrinate also mediated another photoelectrosynthesis for the formation of fine chemicals. The electrolysis of bromoalkyl acrylates was carried out in DMF in the presence of a catalytic amount of heptamethyl cobyrinate (Equation 3.9) [17]. A series of 6-, 10-, and 16-membered cyclic lactones was obtained during the photoirradiation (tungsten lamp, 500 W). These macrocyclic lactones are used in the pharmaceutical industry. In this study, a photo-labile alkylated complex was detected by ultraviolet–visible spectroscopy (UV–vis) and electrospray ionization with tandem mass spectrometry (ESI-MS) analyses during electrolysis, and a radical intermediate was trapped by an electron paramagnetic resonance (EPR) spin-trapping technique using α-phenyl *N*-(*t*-butyl)nitrone (PBN) as shown in Figure 3.4.

Such an intramolecular cyclization was also catalyzed by another vitamin B_{12} derivative (see Chart 3.3). The electrolysis of 2-(4-bromobutyl)-2-cyclohexen-1-one during photoirradiation (xenon lamp, 200 W with 350 nm cutoff filter) formed the *cis* and *trans* 1-decalones (Equation 3.10) [18]. In this reaction, the vitamin B_{12} derivative having six carboxylic groups, $B_{12}(\text{COO}^-)_6$, was immobilized on a TiO_2 electrode and was used as the working electrode.

As for the model studies of the coenzyme B_{12} catalyzed methylmalonyl to succinyl rearrangement, the interaction between a vitamin B_{12} derivative containing a peripheral C_{18} alkyl chain and a (methyl)thiomalonate substrate bearing long alkyl chains at the thioester group was effectively used for electrolysis during the photoirradiation as shown in Figure 3.5. The electrolysis of thioesters at −0.85 V vs. SCE in MeOH/H_2O (v/v, 4:1)

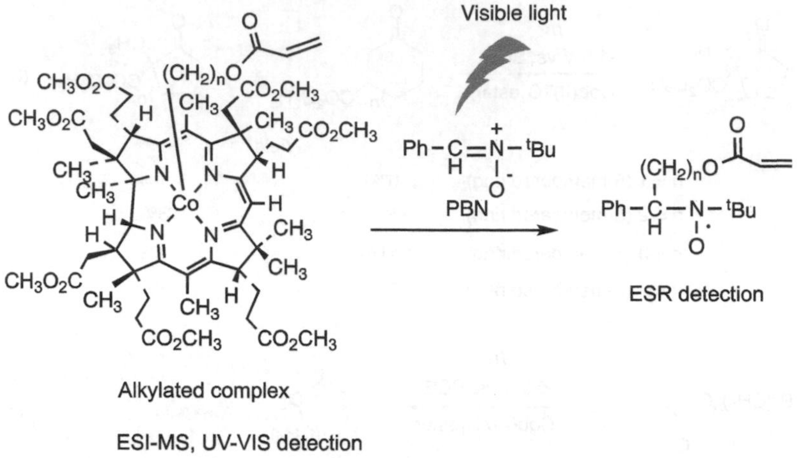

FIGURE 3.4 Detection of photosensitive alkylated complex and radical species.

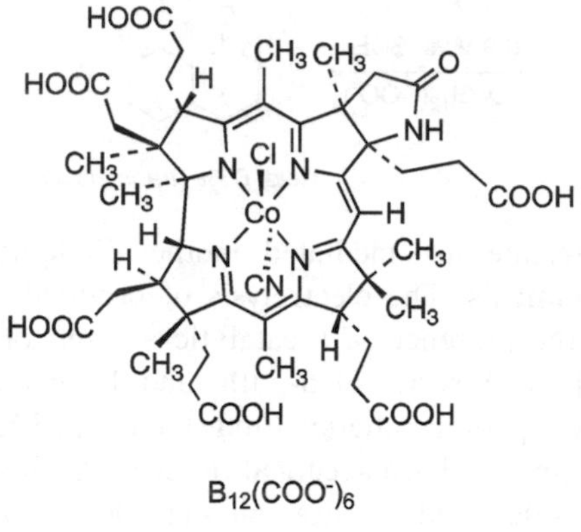

$B_{12}(COO^-)_6$

CHART 3.3

in the presence of the 5% vitamin B_{12} derivative, hexamethyl *c*-octadecyl cobyrinate, during photoirradiation afforded a 22% rearranged product [19,20]. The absence of a long alkyl chain in the catalyst, i.e., heptamethyl cobyrinate ($Cob(II)7C_1ester$), results in a sluggish reaction with a low yield and poor reproducibility. Noncovalent associations between the substrate and the catalyst is essential for this reaction.

3.5 LIGHT-ASSISTED ELECTROSYNTHESIS BY SIMPLE VITAMIN B_{12} MODEL COMPLEX

Cobalt complexes, except for cobalamin, have been synthesized and utilized for the model reaction of the B_{12}-dependent enzyme. Tetradentate chelate compounds, such as porphyrin, phthalocyanine, and Schiff-base compounds, were extensively studied

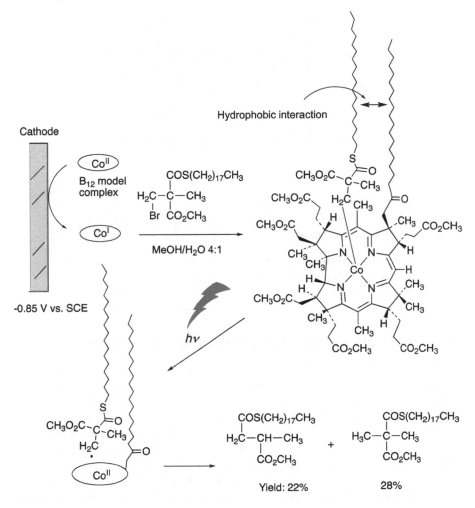

FIGURE 3.5 Effect of hydrophobic peripheral association in alkylated complex during electrolysis and irradiation (MeOH/H$_2$O 4:1).

as ligand for cobalt complex. Among the model complexes, the imine/oxime-type complex, [Co(DH)$_2$(H$_2$O)Cl] and [Co(DO)(DOH)pnBr$_2$] as shown in Chart 3.4 were used as the mediator in the light-assisted electrolysis. For example, a ring-expansion reaction via a 1,2-acyl migration proceeded by the electrolysis of 2-alkyl-2-(bromomethyl)cycloalkanones with [Co(DH)$_2$(H$_2$O)Cl].

The ring-expanded 3-alkyl-2-alkenones were obtained by constant current electrolysis at 20 mA cm^{-2} during photoirradiation (tungsten lamp, 750 W) in moderate yields (Equation 3.11) [21]. The same reaction in Equation 3.6 was also mediated by the simple vitamin B$_{12}$ model complex, [Co(DO)(DOH)pnBr$_2$], but the yield of 2,2-bis(ethoxycarbonyl)propane was low [22,23].

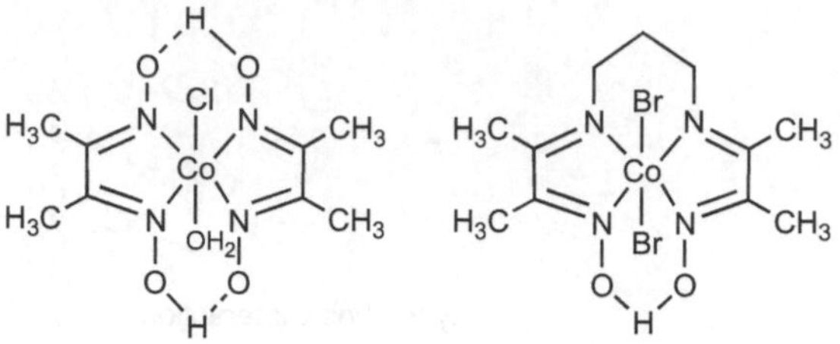

[Co(DH)₂(H₂O)Cl]   [Co(DO)(DOH)pnBr₂]

CHART 3.4

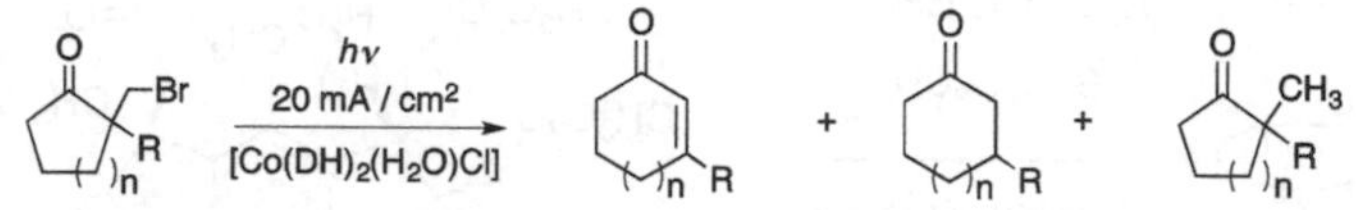

R = n-C₆H₁₃, n = 1 (5-membered ring)   74%   4%   17%
R = n-C₄H₉, n = 1 (5-membered ring)   52%   3%   15%
R = n-C₁₁H₂₃, n = 1 (5-membered ring)   69%   2%   18%
R = n-C₆H₁₃, n = 2 (6-membered ring)   51%   trace   23%

3.6 OUTER OR INNER PHOTOVOLTAIC DEVICE SYSTEMS

Photo-irradiation is also utilized to obtain electricity for electroorganic synthesis. Especially, sunlight is an abundant and sustainable light source and is applied to various electrochemical reactions using a photovoltaic cell [24]. A simple setup with a photovoltaic power supply (*outer photovoltaic device*), readily available as an inexpensive cell, was used to conduct the electrochemical oxidations as shown by Equations 3.12–3.15. Products were formed in good yields comparable to the reactions conducted with the traditional electrochemical setup. This system was also useful for the recycling of Os(VIII)-, TEMPO-, Ce(IV)-, Pd(II)-, Ru(VIII)-, and Mn(V)-oxidants by indirect electrolyses [25]. The versatility of the method demonstrated here clearly shows the environmental advantage of electrochemical reactions combined with photo-energy.

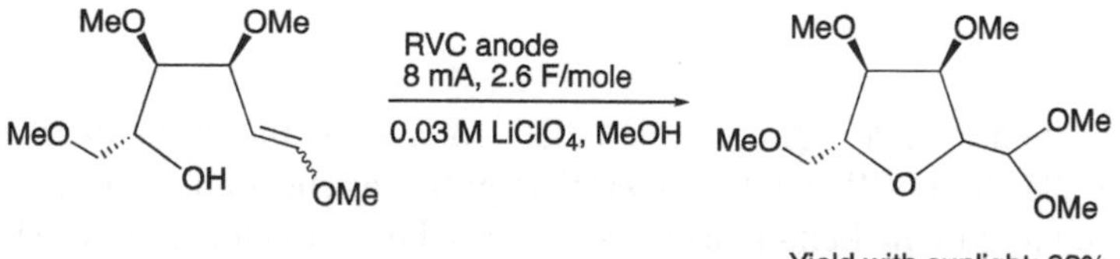

Yield with sunlight: 82%

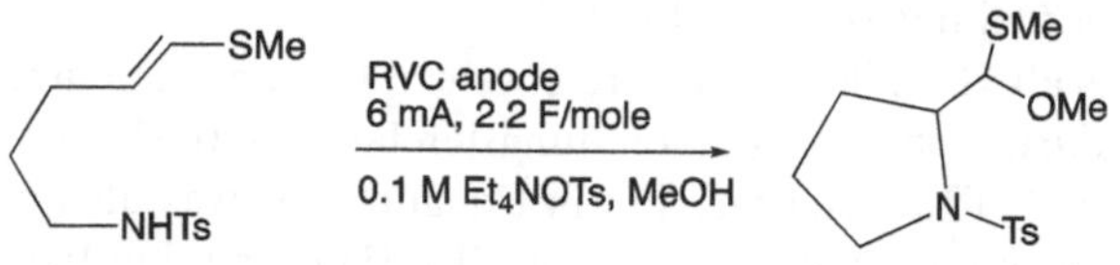

Yield with sunlight: 84%

$$\text{(14)}$$

Yield with sunlight: 52%   14%

$$\text{(15)}$$

Yield with sunlight: 81%

Solar cells inspired by photosynthesis have been developed for decades. Direct modification of a molecular catalyst on the electrode (*inner photovoltaic device*) enables water splitting by photoirradiation [26,27]. For example, the molecular device composed of water oxidation catalyst and photosensitizer, both mainly used in a ruthenium complex, co-adsorbed on a nanostructured TiO_2 allows for visible light-driven water oxidation to form oxygen while hydrogen evolution occurred on the counter Pt electrode as shown in Figure 3.6 [28].

By applying a 0.2 V external bias vs. NHE, a high photocurrent density of more than 1.7 mA cm^{-2} has been achieved. Besides the ruthenium complex, a manganese complex covalently attached to a TiO_2 electrode via a light-absorbing organic linker was used in the photooxidation of a fluorescein derivative to form the fluorescent

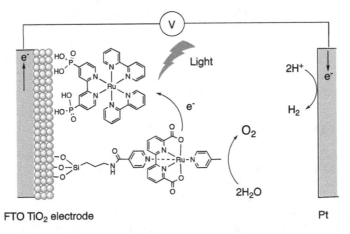

FIGURE 3.6 Photoelectrochemical water oxidation using Ru catalyst. FTO, fluorine-doped tin oxide.

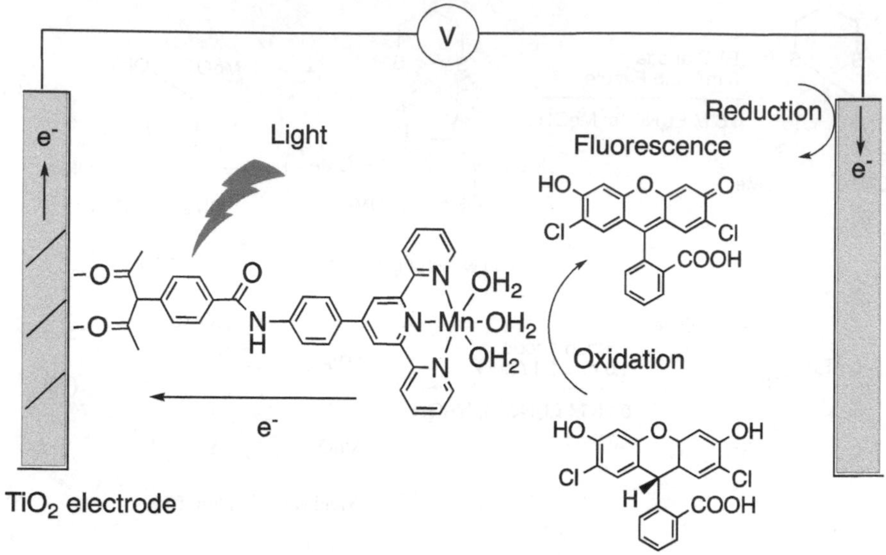

FIGURE 3.7   Photoelectrochemical oxidation of substrate using a surface Mn catalyst.

compound as shown in Figure 3.7 [29]. The photoelectrochemical oxidation of an organic compound as well as water should open the door for the application of solar cells in various molecular transformations.

3.7   CONCLUDING REMARKS

Photoelectrochemistry is progressing more than ever due to its expanding advantage as a clean and economical process as well as producing new photoorganic reactions. One of the future directions of photoelectrochemistry is the smart design of the illuminated cell. Efficient photoirradiation of an electrolyte solution is important for increasing the reaction yield. For example, the recent development of micro-reactor technology in electrochemistry could be a promising approach in pho-toelectrochemistry due to its intelligent design, suitable for photoirradiation [30]. Photoelectrochemically driven electrosynthesis would be developed for versatile applications regarding organic synthesis and energy production.

REFERENCES

1. Francke R.; Little, R.D. 2014. Redox catalysis in organic electrosynthesis: Basic princi-ples and recent developments. *Chem. Soc. Rev.*, 43:2492–2521.
2. Yoshida, J.; Kataoka, K.; Horcajada, R.; Nagaki, A. 2008. Modern strategies in electro-organic synthesis. *Chem. Rev.*, 108:2265.
3. Hisaeda, Y.; Nishioka, T.; Inoue, Y.; Asada, K.; Hayashi, T. 2000. Electrochemical reactions mediated by vitamin B_{12} derivatives in organic solvents. *Coord. Chem. Rev.*, 198:21–37.

4. Kräutler, B. 1999. B_{12} electrochemistry and organometallic electrochemical synthesis. In *Chemistry and Biochemistry of B_{12}*, John Wiley, New York, 315–339.
5. Scheffold, R. 1983. *Modern Synthetic Methods*, Vol. 3, Wiley, Frankfurt, Germany.
6. Scheffold, R.; Orlinski, R. 1983. Carbon–carbon bond formation by light-assisted B_{12} catalysis. Nucleophilic acylation of Michael olefins. *J. Am. Chem. Soc.*, 105:7200–7202.
7. Walder, L.; Orlinski, R. 1987. Mechanism of the light-assisted nucleophilic acylation of activated olefins catalyzed by vitamin B_{12}. *Organometallics*, 6:1606–1613.
8. Lee, E.R.; Lakomy, I.; Bigler, P.; Scheffold, R. 1991. Reductive radical cyclizations of bromo acetals and (bromomethyl)silyl ethers of terpenoid alcohols. *Helv. Chim. Acta*, 74:146–162.
9. Busato, S.; Scheffold, R. 1994. Vitamin B_{12}-catalyzed C,C-bond formations: Synthesis of Jasmonates via sequential radical reaction. *Helv. Chim. Acta*, 77:92–99.
10. Murakami, Y.; Hisaeda, Y.; Kajihara, A. 1984. Hydrophobic vitamin B_{12}. II. Coordination geometry and redox behavior of heptamethyl cobyrinate in nonaqueous media. *Bull. Chem. Soc. Jpn.*, 57:405–411.
11. Murakami, Y.; Hisaeda, Y.; Tashiro, T.; Matsuda, Y. 1985. Electrochemical carbon-skeleton rearrangement as catalyzed by hydrophobic vitamin B_{12} in nonaqueous media. *Chem. Lett.*, 14:1813–1816.
12. Murakami, Y.; Hisaeda, Y.; Tashiro, T.; Matsuda, Y. 1986. Reaction mechanisms for electrochemical carbon-skeleton rearrangement as catalyzed by hydrophobic vitamin B_{12} in nonaqueous media. *Chem. Lett.*, 15:555–558.
13. Banerjee, R.; Ragsdale, S.W. 2003. The many faces of vitamin B_{12}: Catalysis by cobalamin-dependent enzymes. *Ann. Rev. Biochem.*, 72:209–247.
14. Brown, K.L. 2005. Chemistry and enzymology of vitamin B_{12}. *Chem. Rev.*, 105:2075–2149.
15. Murakami, Y.; Hisaeda, Y.; Ozaki, T.; Tashiro, T.; Ohno, T.; Tani, Y.; Matsuda, Y. 1987. Hydrophobic vitamin B_{12}: 8. Carbon-skeleton rearrangement reactions catalyzed by hydrophobic vitamin B_{12} in octopus azaparacyclophane. *Bull. Chem. Soc. Jpn.*, 60:311–324.
16. Hisaeda, Y.; Takenaka, J.; Murakami, Y. 1997. Hydrophobic vitamin B_{12}. Part 14. Ring-expansion reactions catalyzed by hydrophobic vitamin B_{12} under electrochemical conditions in nonaqueous medium. *Electrochim. Acta*, 42:2165–2172.
17. Shimakoshi, H.; Nakazato, A.; Hayashi, T.; Tachi, Y.; Naruta, Y.; Hisaeda, Y. 2001. Electroorganic syntheses of macrocyclic lactones mediated by vitamin B_{12} model complexes: Part 17. hydrophobic vitamin B_{12}. *J. Electroanal. Chem.*, 507:170–176.
18. Mbindyo, J.K.N.; Rusling, J.F. 1998. Catalytic electrochemical synthesis using nanocrystalline titanium dioxide cathodes in microemulsions. *Langmuir*, 14:7027–7033.
19. Wolleb-Gygi, A.; Darbre, T.; Siljegovic, V.; Keese, R. 1994. The importance of peripheral association for vitamin B_{12} catalyzed methylmalonyl-succinyl-rearrangement. *Chem. Commun.*, 835–836.
20. Darbre, T.; Keese, R.; Siljegovic, V.; Wolleb-Gygi, A. 1996. Model studies for the coenzyme-B_{12}-catalyzed methylmalonyl-succinyl rearrangement. The importance of hydrophobic peripheral associations. *Helv. Chim. Acta*, 79:2100–2113.
21. Inokuchi, T.; Tsuji, N.; Kawafuchi, H.; Torii, S. 1991. Indirect electroreduction of 2-alkyl- 2-(bromomethyl)cycloalkanones with cobaloxime to form 3-alkyl-2-alkenones via 1,2-acyl migration. *J. Org. Chem.*, 56:5945–5948.
22. Murakami, Y.; Hisaeda, Y.; Fan, S.D. 1987. Characterization of a vitamin B_{12} model complex and its catalysis in electrochemical carbon-skeleton rearrangement. *Chem. Lett.*, 16:655–658.

23. Murakami, Y.; Hisaeda, Y.; Fan, S.D.; Matsuda, Y. 1989. Redox behavior of simple vitamin B_{12} model complexes and electrochemical catalysis of carbon-skeleton rearrangements. *Bull. Chem. Soc. Jpn.*, 62:2219–2228.
24. Anderson, L.A.; Redden, A.; Moeller, K.D. 2011. Connecting the dots: Using sunlight to drive electrochemical oxidations. *Green Chem.*, 13:1652–1654.
25. Nguyen, B.H.; Redden, A.; Moeller, K.D. 2014. Sunlight, electrochemistry, and sustainable oxidation reactions. *Green Chem.*, 16:69–72.
26. Duan, L.; Tong, L.; Xu, Y.; Sun, L. 2011. Visible light-driven water oxidation-from molecular catalysts to photoelectrochemical cells. *Energy Environ. Sci.*, 4:3296–3313.
27. Concepcion, J.J.; Jurss, J.W.; Brennaman, M.K.; Hoertz, P.G.; Patrocinio, A.O.T.; Iha, N.Y.M.; Templeton, J.L.; Meyer, T.J. 2009. Making oxygen with ruthenium complexes. *Acc. Chem. Res.*, 42:1954–1965.
28. Gao, Y.; Ding, X.; Liu, J.; Wang, L.; Lu, Z.; Li, L.; Sun, L. 2013. Visible light driven water splitting in a molecular device with unprecedentedly high photocurrent density. *J. Am. Chem. Soc.*, 135:4219–4222.
29. Durrell, A.C.; Li, G.; Koepf, M.; Young, K.J.; Negre, C.F.A.; Allen, L.J.; McNamara, W.R.; Song, H.; Batista, V.S.; Crabtree, R.H.; Brudvig, G.W. 2014. Photoelectrochemical oxidation of a turn-on fluorescent probe mediated by a surface Mn^{II} catalyst covalently attached to TiO_2 nanoparticles. *J. Catal.*, 310:37–44.
30. Atobe, M. 2017. Organic electrosynthesis in flow microreactor. *Curr. Opin. Electrochem.* 2:1–6.

4 Ultrasound Activation in Electrosynthesis

Mahito Atobe
Yokohama National University

Frank Marken
University of Bath

CONTENTS

4.1 Introduction 81
4.2 Experimental Methods in Sonoelectrosyntheses... 82
4.3 Mass Transfer Promotion in Electrosynthetic Processes under
    Ultrasonication.. 85
4.4 Product Selectivity Control in Electrosynthetic Processes under
    Ultrasonication.. 87
4.5 Ultrasonic Dispersion in Multiphase Electrolyte Systems 89
4.6 Ultrasound in Electropolymerization 91
4.7 Concluding Remarks 92
References.. 92

4.1 INTRODUCTION

The many benefits of power ultrasound in chemical processes are well documented [1–3] and have been investigated and exploited in a variety of fields in chemical synthesis including electrosynthesis. The most striking effects of ultrasound are observed in heterogeneous reaction systems, particularly those with a solid–liquid interfaces where particle size modification, the modification of particle dispersion, the enhancement of mass transport, the cleaning/activation of surfaces (mainly due to interfacial cavitation), or the formation of fresh reactive surfaces are desirable to promote processes [4].

Because electrochemical reactions are typically heterogeneous in nature with processes confined to the solid (electrode)lliquid (electrolytic solution) interface, various effects of ultrasound, particularly promotion of mass transport, would be induced by ultrasonication. The work of Moriguchi in 1934 was the first example investigating electrochemical processes under ultrasonication [5]. In this study, it was demonstrated that ultrasonication reduced the decomposition voltage of water at platinum. After that, many interesting studies on ultrasonic effects in a variety of electrochemical fields such as electroanalysis, electrosynthesis, and electroplating have been reported [6]. This research field has now become known as "sonoelectrochemistry" [7,8]. Although interest in ultrasound methods in electrochemistry emerged in the 1950s [9], the application of ultrasound to electrosynthesis had been very rarely studied before the 1980s. Since then, there has been a notable increase in productivity in this field, and this area has now become an exciting topic of sonoelectrochemical research where the fundamental effects of ultrasound as a function of frequency are exploited to improve/enable otherwise difficult electrosynthetic processes.

Given the fact that power ultrasound substantially increases mass transport, it is now possible to more effectively electrolyze slowly diffusing clusters or particles. A new development more recently is the application of "impact processes" where solid particles in suspension are ultrasonically accelerated onto the electrode surface to undergo electrolysis. This can lead to entirely new types of electrode reactions based on nanoparticles and nanomaterials. Processes involving "slow diffusers" [10] such as colloids or nanoparticles are most affected by the application of power ultrasound. This is the reason this chapter is dedicated to the innovative technique called "sonoelectrosynthesis" and to its contribution to the improvement of electrosynthetic processes.

4.2 EXPERIMENTAL METHODS IN SONOELECTROSYNTHESES

A number of experimental arrangements have been used for the introduction of ultrasound into an electrosynthetic system. The most straight forward way of sonoelectrosynthesis can be realized by immersion of an electrochemical cell into an ultrasonic cleaning bath. This is illustrated in Figure 4.1, and the results of the electrochemical reduction of

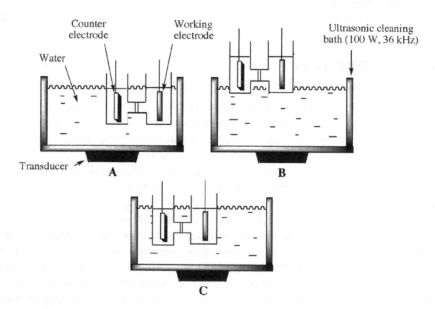

FIGURE 4.1 (A–C) Placement of cells and electrodes in an ultrasonic cleaning bath (100 W, 36 kHz). (From Atobe, M. and Nonaka, T., *Nippon Kagaku Kaishi*, 219, 1998. With permission.)

TABLE 4.1
Electroreduction of Benzaldehyde (40 mM)[a] in an Ultrasonic Cleaning Bath (100 W, 36 kHz)

$$ArCHO \xrightarrow{e^- + H^+/ 2e^- + 2H^+} \tfrac{1}{2} \underset{\overset{|}{OH}\ \overset{|}{OH}}{ArCH-CHAr} \Big/ ArCH_2OH$$

(D1)   (M1)

Stirring Mode	Current Efficiency for [D1]+[M1]/%	Product Selectivity for [D1] /%
Still	36	40
Mechanical[b]	63	72
Ultrasonic A[c]	66	72
Ultrasonic B[c]	70	96
Ultrasonic C[c]	69	97

[a] Electrolyzed by passing 0.5 F mol^{-1} at 20 mA cm^{-2} on a lead cathode in a 0.25 M H$_2$SO$_4$/50% MeOH solution.

[b] Stirred by a rotating magnet bar.

[c] See Figure 4.1.

benzaldehyde in the H-type divided cell located in different places within the ultrasonic bath are shown in Table 4.1 [11]. This method is simple, and significant ultrasonic effects on current efficiency and product selectivity can be obtained. However, mass transport calibration methods (as well as thermal calibration of energy transfer) should be employed. Otherwise, this method can be irreproducible as the results are strongly dependent on the precise placement of the cell within the bath, as shown in Table 4.1. In addition, the specific geometry of the cell used also significantly affects the electrolysis in this method.

A much improved cell design with a sonic probe in a separate compartment was reported by Degrand and coworkers [12] who successfully electrogenerated mono- and di-telluride and selenide anions. An even more direct method by which ultrasound can be introduced into an electrochemical system is by using an ultrasonic immersion probe. This can be a glass immersion horn [13] or, with appropriate precautions in experimental design, a metal immersion horn [14] based typically on mechanically robust Ti$_6$Al$_4$V alloy. An ultrasonic probe consists of a transducer driving an acoustic horn oscillator, which transmits the vibrational energy from its tip into the solution environment. In the most used configuration, both the electrodes and the probe are immersed in the same electrolyte solution with the horn tip facing the working electrode at a known distance, as shown in Figure 4.2. The ultrasonic probe system has several advantages over an ultrasonic bath for sonoelectrosynthesis. Because the ultrasound power is also easily controlled in ultrasonic probe system, the effects seen are directional and reproducible and easily interpreted. In addition, the acoustic power applied by the probe system can be significantly higher, and hence comparable/improved results to those using the bath system can be obtained with less power (see Table 4.2) [11].

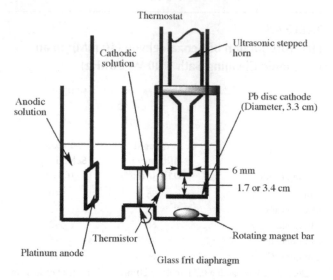

FIGURE 4.2 Electrolytic cell with an ultrasonic horn (a titanium alloy rod with 6 mm diameter, 0–20 W, 20 kHz). (From Atobe, M. and Nonaka, T., *Nippon Kagaku Kaishi*, 219, 1998. With permission.)

TABLE 4.2
Electroreduction of Benzaldehyde (40 mM)[a] under Ultrasonication with a Stepped Horn

$$\text{ArCHO} \xrightarrow{\text{e}^- + \text{H}^+/\, 2\text{e}^- + 2\text{H}^+} \tfrac{1}{2}\ \underset{\substack{| \quad |\\ \text{OH} \quad \text{OH}}}{\text{ArCH}-\text{CHAr}} \Big/ \text{ArCH}_2\text{OH}$$

(D1)   (M1)

Ultrasonic Power/W	Distance Between Horn Tip and Electrode/cm	Current Efficiency for [D1]+[M1]/%	Product Selectivity for [D1]/%
4	1.7	72	93
8	1.7	80	96
12	1.7	89	98
12	3.4	70	94

[a] Electrolyzed by passing 0.5 F mol^{-1} at 20 mA cm^{-2} on a lead cathode in a 0.25 M H$_2$SO$_4$/50% MeOH solution.

Another method proposed for high-energy applications is based on the so-called "sonotrode" geometry (Figure 4.3) [15]. This is when the tip of the horn is used as the electrode or is modified such that an electrode is physically introduced into the tip of the horn. The sonotrode geometry has many different characteristics compared to the face-on geometry illustrated in Figure 4.2. As well as causing cavitational cleaning, the vigorous shaking of the horn can enhance the detachment of bulk

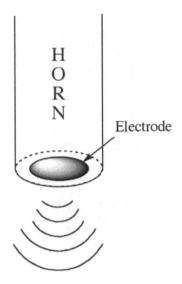

FIGURE 4.3 A schematic diagram showing the "sonotrode." (From Marken, F. et al., *J. Electroanal. Chem.*, 415, 55, 1996. With Permission.)

electrodeposited surface species. Extreme levels of mass transport with diffusion layer thicknesses less than 1 μm are accessible [16], which translates into high-speed electrolysis.

4.3 MASS TRANSFER PROMOTION IN ELECTROSYNTHETIC PROCESSES UNDER ULTRASONICATION

The description of mass transport phenomena in electrochemical processes under ultrasonication has been the subject of many papers and a stepwise development of better understanding. Three major contributions to the mass transport were (i) macroscopic acoustic streaming (Figure 4.4a) and more localized (ii) microjetting (Figure 4.4b) resulting from cavitational collapse of bubbles near the electrode surface [17]. In addition, (iii) interfacial bubble oscillations have been observed to exert strong forces on the electrode surface [18].

These contributions can be confirmed as an increase in the limiting current in the voltammetric study. More in-depth study of the microjetting phenomenon revealed that also repeated bubble expansion and contraction close to or directly on the electrode surface can have major mechanical effects [18] adding to the conventional microjetting phenomenon.

In the case for the preparative electrosynthesis, the overall mass transfer promotion by ultrasonication provides an increase in the current efficiency and/or the yield. For example, as mentioned in the former section, a significant ultrasonic effect on the current efficiency was found in the electrochemical reduction of benzaldehyde (see Tables 4.1 and 4.2) [11]. The current efficiency for the reduction of benzaldehyde was dramatically increased under ultrasonication, and furthermore, the product selectivity for the hydrodimeric product (HD) was also increased by ultrasonication,

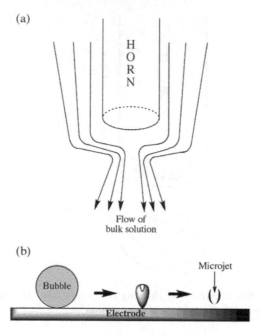

FIGURE 4.4 Schematic showing (a) the acoustic streaming from an ultrasonic stepped horn and (b) the formation of a microjet from the asymmetric collapse of a cavitation bubble near the electrode. (Adapted with permission from Maisonhaute, E.; White, P.C.; Compton, R.G. 2001. Surface acoustic cavitation understood via nanosecond electrochemistry. *J. Phys. Chem. B* 105:12087–12091. Copyright 2001 American Chemical Society.)

and the effects could be rationalized experimentally and theoretically as due to the promotion of mass transport of the substrate molecule to the electrode surface from the electrolytic solution by ultrasonication [19].

Compton and coworkers successfully demonstrated an improved current efficiency (3 F for a direct two-electron reduction) and a clean conversion (yields up to 80% isolated) in the reductive ring opening of the α,β-epoxyketone isophorone oxide to yield the corresponding β-hydroxyketone (Scheme 4.1) [20]. They attributed the considerably enhanced performance and current efficiency in electrosynthesis to selectively increasing the rate of the mass transport-controlled processes achieved in the presence of ultrasound.

Cavitation effects and microjetting can also introduce detrimental effects, in particular when working electrode materials are prone to embrittlement and crack formation, e.g., glassy carbon [21]. Therefore, novel electrode materials such as

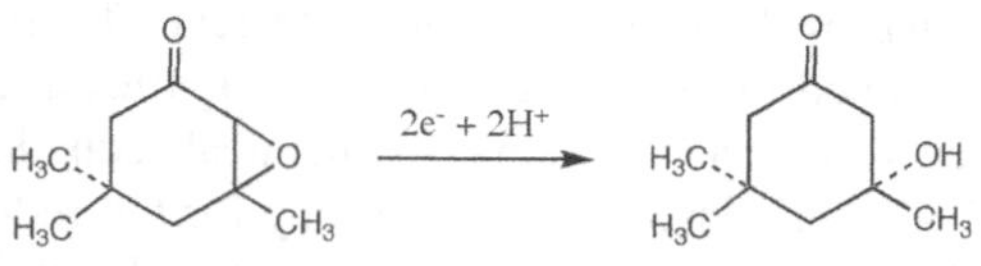

SCHEME 4.1 Reductive ring opening of the α,β-epoxyketone isophorone oxide.

boron-doped diamond can be very helpful and remain operationally stable even after prolonged use with power ultrasound or under much more extreme conditions [22].

4.4 PRODUCT SELECTIVITY CONTROL IN ELECTROSYNTHETIC PROCESSES UNDER ULTRASONICATION

Ultrasound can often alter the course of the reaction and thereby control the product selectivity in electrosynthetic processes. Early studies of this effect were performed by the Coventry Group in the late 1980s [23]. As a model process, they chose the Kolbe reaction (the electrooxidation of carboxylate anions), which has long been known as one of the earliest discovered electroorganic reactions. Different reaction pathways exist under different kinetic regimes, and adsorption and other electrode phenomena are known to be important. Hence, all pathways would be influenced to differing extents by ultrasonication. Scheme 4.2 shows plausible reaction scheme of the electrooxidation of a typical carboxylate anion ($RCOO^-$). In general, the mechanism breaks down into a pathway involving one electron per molecule of starting material, giving products from the radical intermediate, for example the dimmer (R-R, the Kolbe reaction product), and a two-electron pathway per starting molecule, giving products from an intermediate cation (the Hofer-Moest reaction product).

Table 4.3 shows product ratios from the electrooxidation of partially neutralized cyclohexanecarboxylate at platinum anode in a methanol electrolytic solution. In the absence of ultrasound, the results show a substantial amount (49%) of the dimer bicyclohexyl from the one-electron pathway, together with cyclohexylmethyl ester, cyclohexanol, and other products from the two-electron pathway (approx. 30%). In sharp contrast, in the presence of ultrasound, there is only 8% of the bicyclohexyl dimeric one-electron product, with approximately 41% of the two-electron product

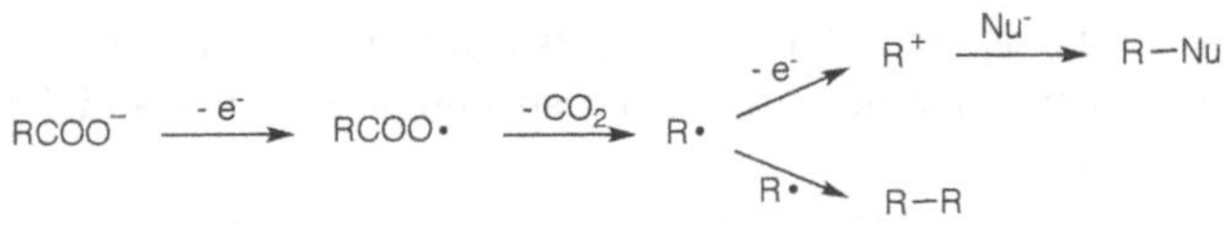

SCHEME 4.2   Electrooxidation of a typical carboxylate anion ($RCOO^-$).

TABLE 4.3
Product Ratios from the Electrooxidation of Partially Neutralized Cyclohexane-Carboxylate at a Platinum Anode in a Methanol Electrolytic Solution

Reaction Conditions	Bicyclohexyl	Cyclo-hexane	Cyclo-hexane	Methoxy-cyclohexane	Methyl-cyclohexane	Cyclo-hexanol
Silent	49	1.5	4.5	24.9	17	2.1
Ultrasound	7.7	2.6	32.4	34.3	2.5	6.8

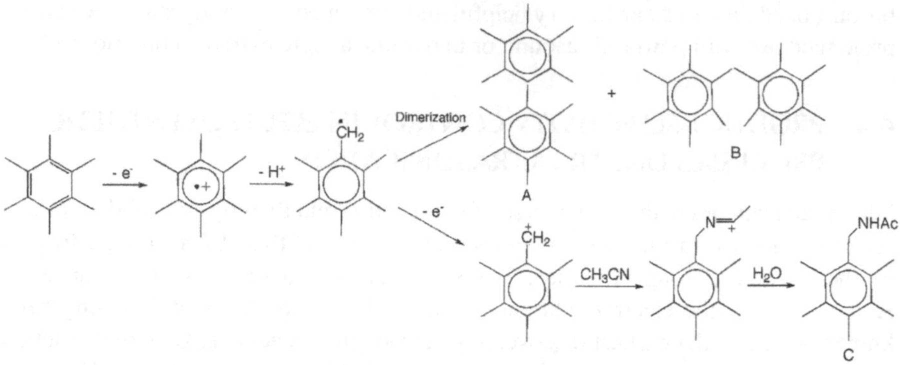

SCHEME 4.3   Electrochemical oxidation of pentamethylbenzene.

TABLE 4.4

Product Ratios for the Electrochemical Oxidation of Pentamethylbenzene

Reaction Conditions	A	B	C
Mechanical stirring	8	14	42
Ultrasound (38 kHz)	9	14	62
Ultrasound (800 kHz)	7	43	24

from nucleophilic capture of the intermediate carbocation. The preponderance of cyclohexene (32%) over cyclohexane (> 3%) indicates its formation is by proton loss from the carbocation intermediate. From these facts, it can be stated that the condition under ultrasound appeared to favor the two-electron mechanism, and the greatest effect of ultrasonication upon product distribution was substantial enhancement of alkene formation.

Another example where the frequency of ultrasonication is important is in the electrooxidation of alkylaromatics [24]. In this oxidation, a number of possible products are obtained, and the product ratio depends on the substrate and the reaction conditions. As shown in Scheme 4.3, for pentamethylbenzene, the dimerization occurs and gives the products A and B. In addition, an acetoamidation (the nucleophilic substitution on the benzyl position with acetonitrile) also occurs competitively to give the amide C in an acetonitrile solution. Table 4.4 shows the percentage product ratio in the electrooxidation of pentamethylbenzene for stirred and sonicated conditions. The amide C was the major product in a solution mechanically stirred or sonicated with 38 kHz. However, the use of 800 kHz ultrasound favors diphenylmethane B in a switch to the one-electron mechanism. Thus, sonication at different frequencies, even at equivalent power, has provided different effects upon the reaction outcome.

Low-temperature electrosyntheses are hampered by much lower diffusion coefficients and therefore also benefit from ultrasound activation. The Birch reduction mechanism based on solvated electrons was investigated, and it was suggested that even in the presence of ethanol cosolvent, short-lived solvated electrons can react successfully due to the higher ultrasound-induced rate of mass transport [25,26]. The low-temperature reductive electrodimerization of 2-nitrobenzylchloride in liquid ammonia was demonstrated under power ultrasound conditions [27].

4.5 ULTRASONIC DISPERSION IN MULTIPHASE ELECTROLYTE SYSTEMS

Water is often the most ideal solvent for organic electrosynthesis because of good conductivity and polarity, stabilizing effects on radical intermediates, and straight-forward product isolation by extraction or filtration [28]. However, a number of organic materials cannot be efficiently electrolyzed in aqueous solutions due to their low aqueous solubility. Hence, in most cases, the electrolysis of water-insoluble materials has been carried out in organic polar solvents. However, the electrolysis in organic solvents is practically disadvantageous from the aspects of conductivity, operational costs, and safety. Therefore, the extension of the range of systems under-going an electrochemical synthesis in water is of considerable interest and researches employing surfactant-stabilized emulsions and suspensions have been investigated [29], but the presence of surfactants might cause other problems such as difficulty in their separation from the reaction mixtures and production of wastes. Hence, it is desirable to conduct emulsion or suspension synthetic processes in the absence of surfactants in order to fulfill its potential as a "green" methodology.

On the other hand, it is well known that ultrasound allows multiphase systems (liquid/liquid, liquid/solid, or liquid/gas) to be efficiently homogenized. For example, ultra-sonication provides fast mass transport as well as stable emulsions without stabilizing agents simply by mechanical forces, which arise at the liquid/liquid phase boundaries [30]. This has been termed as "acoustic emulsification" [31]. Furthermore, it is also known that emulsions prepared by acoustic emulsification have the characteristic of giving narrow droplet size distributions in the submicrometer range [32]. Hence, these droplets can be electrolyzed smoothly at electrodes, and therefore, the practical reac-tion rate is obtained even without any emulsifying reagents. Compton and coworkers reported that apparent diffusion rates of water-insoluble molecules (diethylmaleate, diethylfumarate, and diethylacetylene dicarboxylate) in emulsion systems under ultra-sonication were voltammetrically estimated to be large enough to perform prepara-tive electrolyses [33]. For heptane-water emulsions, the mechanism was studied [34] and even the effects of single microdroplet impacts on electrodes have been measured with nanosecond electrochemistry [35]. On the other hand, Atobe et al. reported that electrochemical oxidation of water-insoluble amines (n-octylamine and n-decylamine) was accomplished in aqueous electrolytes using acoustic emulsification and gave much higher current efficiencies compared to those obtained under conditions of mechanical stirring (see Table 4.5) [36]. In addition, they also proposed that the smooth electro-chemical reaction in the emulsions prepared by ultrasonication took place via direct electron transfer between the electrode and the water-insoluble organic droplets.

Suspended particle-electrodes or slurry-electrodes can be used to achieve a high space-time yield since these can play a role as an electron transfer mediator and pos-sess a large surface. Therefore, the application of suspended particle-electrodes to

TABLE 4.5

Current Efficiency in the Anodic Oxidation of Emulsified Amines

$$R\text{-}CH_2\text{-}NH_2 \xrightarrow{\ -4e^-,\ -4H^+\ } R\text{-}C\equiv N \qquad R = C_7H_{15},\ E_{ox} = 0.25\ V\ vs\ SCE$$
$$R = C_9H_{19},\ E_{ox} = 0.25\ V\ vs\ SCE$$

Substrate	Mechanical Stirring/rpm	Ultrasonication/W cm⁻²	Current Efficiency/%
n-Octylamine	500	–	2.5
n-Octylamine	1,000	–	31
n-Octylamine	1,400	–	33
n-Octylamine	–	106	44
n-Octylamine	–	177	68
n-Octylamine	–	248	71
n-Decylamine	1,400	–	17
n-Decylamine	–	248	46

electrosynthetic processes has been studied by many researchers [37–39]. However, the particle-electrode is usually aggregated during electrolysis in the absence of stabilizer, and the total surface area of the electrode is reduced. To overcome this problem, ultrasonic dispersion is very useful for suppression of agglomeration of particle-electrode. Atobe et al. examined ultrasonic effect on the electroreduction of acrylonitrile at suspended lead particle-electrode (see Figure 4.5) [40,41]. The product selectivity for adiponitrile **1**, which was formed as the corresponding HD along with propionitrile **2** as the hydromonomeric one in the cathodic reduction of acrylonitrile, was increased by the addition of lead particles as a particle-electrode, and, moreover, the selectivity was further increased under ultrasonication. They concluded that ultrasonic effect was ascribed to not only mass transport of lead

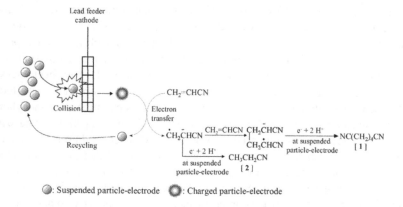

FIGURE 4.5 Schematic representation of the electroreduction of acrylonitrile at suspended lead particle-electrode. (Asami, R. et al., *Ultrason. Sonochem.*, 13, 19, 2006. With permission.)

particles to the feeder cathode but also an increase in the effective surface area of particle-electrode by ultrasonic dispersion.

Direct interaction of microparticles [42] and nanoparticles [43] can be enhanced and studied in the presence of ultrasound. Ultrasound-enhanced "impact processes" have been reported, for example, for Fe_2O_3 [44] and graphite powder [45]. For these solid particles, collision rates with the electrode and transport rates are enhanced and quantitative conversion may be feasible with electrosynthesis providing a new way for clean solid–solid transformation. Sonoelectrochemical reactors for the electrosynthesis of particulate products have been proposed [46].

4.6 ULTRASOUND IN ELECTROPOLYMERIZATION

A particularly important case of electroorganic synthesis is electropolymerization. Over the last couple of decades, conducting polymers have attracted considerable attention for a wide variety of potential technological applications. The uniquely switchable electrical, optical, and chemical properties offered by these materials has been reported to be beneficial in anticorrosion coatings [47], electromagnetic shielding [48], rechargeable polymeric batteries [49], polymer photovoltaics [50], polymer actuators [51,52], and so on. However, a major problem in their successful utilization is their poor mechanical properties and processability due to their insoluble nature in common organic solvents. Incorporation of polar functional groups or long and flexible alkyl chains in the polymer backbone is a common technique to prepare conducting polymers, which are soluble in water and/or organic solvents [53]. On the other hand, structure control during synthesis of conducting polymer materials (e.g., by application of ultrasound) is another approach to overcome poor processability. Therefore, many methods for controlling their physical structures have been reported so far. Solvent, concentrations, temperature, and additives have all been shown to play a role for this purpose [54]. Furthermore, in recent years, an increase in interest has been observed in controlling polymer morphological properties using original techniques such as the use of ultrasound in polymerization processes, particularly electropolymerization [55].

Ultrasound offers a powerful tool in electropolymerization [56]. Osawa et al. have found that the quality of polythiophene films electropolymerized on an anode can be enhanced by ultrasound [57]. By conventional methodology, the films become brittle, but by using ultrasound from a 45 kHz cleaning bath, flexible and tough films (tensile modulus 3.2 GPa and strength 90 MPa) can be obtained. Work by Atobe and coworkers was the first systematic investigation of electropolymerization under sonication in a complete series of papers on work performed at lower frequency ultrasound [58–60]. The behavior of polypyrrole films electropolymerized under ultrasonication was investigated, and unique properties in the doping–undoping processes were highlighted. The authors attributed these results to the development of highly dense films under sonication, but also observed the degradation of the film due to higher cavitational damage at 20 kHz [61]. The Besançon group and, in particular, Taouil et al. studied the use of high-frequency ultrasound (500 kHz, 25 W) for electropolymerization of 3,4-ethylenedioxythiophene or polypyrrole in aqueous medium in order to investigate its effects on conducting polymer properties. They showed that (i) mass transfer enhancement induced by sonication improves electropolymerization and that (ii) mass transfer effect is not the only phenomenon induced by ultrasound during electrodeposition [62,63]. The effect of standing ultrasonic pressure waves imprinting a pattern into electropolymers was reported by Walton and coworkers [64].

4.7 CONCLUDING REMARKS

Many studies on ultrasonic effects in a variety of electrochemical fields have been reported since the pioneering work of sonoelectrochemistry by Moriguchi in 1934, while application of ultrasound to electrosynthesis had been very little studied for many years. There has been a gradual upsurge of interest since the late 1980s, and now this area has become an exciting topic of sonoelectrochemical research, because the use of ultrasound offers a number of significant benefits in laboratory-scale electrosynthetic processes including improved current yields and cleaner conversion with products kinetically selected by mass transport control. Particular benefits are observed in low-temperature electrosynthesis and in processes with "slow diffusers" such as those based on colloids and particles.

The ability to explore redox reactions at the level of single microdroplets during impact at the electrode surface substantially improves the methodology for mechanistic studies, and this will lead to better processes. We believe that, in future, ultrasound-enhanced electrosynthesis will be developed further and successfully applied to a wider spectrum of electrosynthetic reactions. New types of innovative ultrasonic reactors will be needed for scale-up of processes.

REFERENCES

1. Suslick, K. 1988. *Ultrasound: Its Chemical, Physical, and Biological Effects*. VCH, Weinheim.
2. Nowak, F.M. 2011. *Sonochemistry: Theory, Reactions & Syntheses, & Applications*. Nova Science Publishers Inc., London.
3. Mason, T.J. 1989. *Sonochemistry: Applications and Uses of Ultrasound in Chemistry*. John Wiley & Sons, London.
4. Walton, D.J.; Phull, S.S. 1996. Sonoelectrochemistry. In Mason T.J. (ed.) *Advances in Sonochemistry*, 205–284. JAI Press, London.
5. Moriguchi, N. 1934. The effect of supersonic waves on chemical phenomena. IV. The effect on the overvoltage. *Chem. Soc. Jpn.* 55:751.
6. Pollet, B.G. 2012. *Power Ultrasound in Electrochemistry*. John Wiley & Sons, London.
7. Yaqub, A.; Ajab, H. 2013. Applications of sonoelectrochemistry in wastewater treatment system. *Rev. Chem. Eng.* 29:123–130.
8. Gonzalez-Garcia, J.; Esclapez, M.D.; Bonete, P.; Hernandez, Y.V.; Garreton, L.G.; Saez, V. 2010. Current topics in sonoelectrochemistry. *Ultrasonics* 50:318–322.
9. Yeager, E.; Hovorka, F. 1953. Ultrasonic waves and electrochemistry. 1. A survey of the electrochemical applications of ultrasonic waves. *J. Acoust. Soc. Am.* 25:443–455.
10. Holt, K.B.; Del Campo, J.; Foord, J.S.; Compton, R.G.; Marken, F. 2001. Sonoelectrochemistry at platinum and boron-doped diamond electrodes: Achieving "fast mass transport" for "slow diffusers." *J. Electroanal. Chem.* 513:94–99.
11. Matsuda, K.; Atobe, M.; Nonaka, T. 1994. Ultrasonic effects on electroorganic processes. Part 1. *Chem. Lett.* 1619–1622.

12. Gautheron, B.; Tainturier, G.; Degrand, C. 1985. Ultrasound-induced electrochemical synthesis of the anions SE_2^{2-}, Se^{2-}, Te_2^{2-}, and Te^{2-}. *J. Am. Chem. Soc.* 107:5579–5581.
13. Holt, K.B.; Sabin, G.; Compton, R.G.; Foord, J.S.; Marken, F. 2002. Reduction of tetrachloroaurate (III) at boron-doped diamond electrodes: Gold deposition versus gold colloid formation. *Electroanalysis* 14:797–803.
14. Marken, F.; Compton, R.G. 1996. Electrochemistry in the presence of ultrasound: The need for bipotentiostatic control in sonovoltammetric experiments. *Ultrason. Sonochem.* 3:S131–S134.
15. Ball, J.C.; Compton, R.G. 1999. Application of ultrasound to electrochemical measurements and analyses. *Electrochemistry* 67:912–919.
16. Marken, F.; Akkermans, R.P.; Compton, R.G. 1996. Voltammetry in the presence of ultrasound: The limit of acoustic streaming induced diffusion layer thinning and the effect of solvent viscosity. *J. Electroanal. Chem.* 415:55–63.
17. Compton, R.G.; Eklund, J.C.; Marken, F. 1997. Dual activation: Coupling ultrasound to electrochemistry: An overview. *Electrochim. Acta* 42:2919–2927.
18. Maisonhaute, E.; Prado, C.; White, P.C.; Compton, R.G. 2002. Surface acoustic cavitation understood via nanosecond electrochemistry. Part III: Shear stress in ultrasonic cleaning. *Ultrason. Sonochem.* 9:297–303.
19. Atobe, M.; Matsuda, K.; Nonaka, T. 1996. Ultrasonic effects on electroorganic processes. Part 4. *Electroanalysis* 8:784–788.
20. Marken, F.; Compton, R.G.; Davies, S.G.; Bull, S.D.; Thiemann, T.; Sá e Melo, M.L.; Neves, A.C.; Castillo, J.; Jung, G.J.; Fontana, A. 1997. Electrolysis in the presence of ultrasound: Cell geometries for the application of extreme rates of mass transfer in electrosynthesis. *J. Chem. Soc., Perkin Trans.* 2:2055–2059.
21. Marken, F.; Kumbhat, S.; Sanders, G.H.W.; Compton, R.G. 1996. Voltammetry in the presence of ultrasound: Surface and solution processes in the sonovoltammetric reduction of nitrobenzene at glassy carbon and gold electrodes. *J. Electroanal. Chem.* 414:95–105.
22. Compton, R.G.; Foord, J.S.; Marken, F. 2003. Electroanalysis at diamond-like and doped-diamond electrodes. *Electroanalysis* 15:1349–1363.
23. Chyla, A.; Lorimer, J.P.; Mason, T.J.; Smith, G.; Walton, D.J. 1989. Modifying effect of ultrasound upon the electrochemical oxidation of cyclohexanecarboxylate. *Chem. Commun.* 9:603–604.
24. Mason, T.J.; Lorimer, J.P. 2002. *Applied Sonochemistry.* Wiley-VCH, Weinheim.
25. Del Campo, F.J.; Neudeck, A.; Compton, R.G.; Marken, F. 1999. Low-temperature sonoelectrochemical processes. Part 1. Mass transport and cavitation effects of 20 kHz ultrasound in liquid ammonia. *J. Electroanal. Chem.* 477:71–78.
26. Del Campo, F.J.; Neudeck, A.; Compton, R.G.; Marken, F.; Bull, S.D.; Davies, S.G. 2001. Low-temperature sonoelectrochemical processes Part 2: Generation of solvated electrons and Birch reduction processes under high mass transport conditions in liquid ammonia. *J. Electroanal. Chem.* 507:144–151.
27. Del Campo, F.J.; Maisonhaute, E.; Compton, R.G.; Marken, F.; Aldaz, A. 2001. Low-temperature sonoelectrochemical processes. Part 3. Electrodimerisation of 2-nitrobenzylchloride in liquid ammonia. *J. Electroanal. Chem.* 506:170–177.
28. Feess, H.; Wendt, H. 1982. Performance of two-phase-electrolyte electrolysis. In Weinberg N.L.; Tilak, B.V. (Eds.) *Technique of Electroorganic Synthesis*, Part III. John Wiley & Sons, New York.
29. Mackay, R.A.; Texter, J. 1992. *Electrochemistry in Colloids and Dispersions.* Wiley-VCH, Weinheim.
30. Banks, C.E.; Rees, N.V.; Compton, R.G. 2002. Sonoelectrochemistry in acoustically emulsified media. *J. Electroanal. Chem.* 535:41–47.

31. Reddy, S.R.; Fogler, H.S. 1980. Emulsion stability of acoustically formed emulsions. *J. Phys. Chem.* 84:1570–1575.
32. Kamogawa, K.; Okudaira, G.; Matsumoto, M.; Sakai, T.; Sakai, H.; Abe, M. 2004. Preparation of oleic acid/water emulsions in surfactant-free condition by sequential processing using midsonic-megasonic waves. *Langmuir* 20:2043–2047.
33. Marken, F.; Compton, R.G.; Bull, S.D.; Davies, S.G. 1997. Ultrasound-assisted electrochemical reduction of emulsions in aqueous media. *Chem. Commun.* 995–996.
34. Banks, C.E.; Rees, N.V.; Compton, R.G. 2002. Sonoelectrochemistry in acoustically emulsified media. *J. Electroanal. Chem.* 535:41–47.
35. Banks, C.E.; Rees, N.V.; Compton, R.G. 2002. Sonoelectrochernistry understood via nanosecond voltammetry: Sono-emulsions and the measurement of the potential of zero charge of a solid electrode. *J. Phys. Chem. B* 106:5810–5813.
36. Atobe, M.; Ikari, S.; Nakabayashi, K.; Amemiya, F.; Fuchigami, T. 2010. Electrochemical reaction of water-insoluble organic droplets in aqueous electrolytes using acoustic emulsification. *Langmuir* 26:9111–9115.
37. Baria, D.N.; Hulburt, H.M. 1973. Reduction of crotonic acid with hydrogen on a slurry electrode. *J. Electrochem. Soc.* 120:1333–1339.
38. Keren, E.; Soffer, A. 1973. Hydrogen adsorption rate and discharge mechanism on palladium-carbon suspension electrode. *J. Electroanal. Chem.* 44:53–62.
39. Kametani, H. 1972. Potential curves in the suspension electrode electrolysis of copper. *Denki Kagaku* 40:449–454.
40. Yoshizawa, S.; Takehara, Z.; Ogumi, Z.; Matsubara, M.; Tsuji, T. 1976. Application of a circulating suspended-particle electrode to electrochemical hydrodimerization of acrylonitrile. *J. Appl. Electrochem.* 6:403–409.
41. Asami, R.; Atobe, M.; Fuchigami, T. 2006. Ultrasonic effects on electroorganic processes. Part 27. Electroreduction of acrylonitrile at suspended lead particle-electrode. *Ultrason. Sonochem.* 13:19–23.
42. Rees, N.V.; Banks, C.E.; Compton, R.G. 2004. Ultrafast chronoamperometry of acoustically agitated solid particulate suspensions: Nonfaradaic and faradaic processes at a polycrystalline gold electrode. *J. Phys. Chem. B* 108:18391–18394.
43. McKenzie, K.J.; Marken, F. 2001. Direct electrochemistry of nanoparticulate Fe_2O_3 in aqueous solution and adsorbed onto tin-doped indium oxide. *Pure Appl. Chem.* 73:1885–1894.
44. Murphy, M.A.; Marken, F.; Mocak, J. 2003. Sonoelectrochemistry of molecular and colloidal redox systems at carbon nanofiber-ceramic composite electrodes. *Electrochim. Acta* 48:3411–3417.
45. Clegg, A.D.; Rees, N.V.; Banks, C.E.; Compton, R.G. 2006. Ultrafast chronoamperometry of single impact events in acoustically agitated solid particulate suspensions. *ChemPhysChem* 7:807–811.
46. Reisse, J.; Francois, H.; Vandercammen, J.; Fabre, O.; Kirschdemesmaeker, A.; Maerschalk, C.; Delplancke, J.L. 1994. Sonoelectrochemistry in aqueous-electrolyte: A new types of sonoelectroreactor. *Electrochim. Acta* 39:37–39.
47. DeBerry, D.W. 1985. Modification of the electrochemical and corrosion behavior of stainless steels with an electroactive coating. *J. Electrochem. Soc.* 132:1022–1026.
48. Ahmed, N.; MacDiarmid, A.G. 1996. Inhibition of corrosion of steels with exploitation of conducting polymers. *Synth. Met.* 78:103–110.
49. Joo, J.; Epstein, A.J. 1994. Electromagnetic radiation shielding by intrinsically conducting polymers. *Appl. Phys. Lett.* 65:2278–2280.

50. MacDiarmid, A.G.; Mu, S.L.; Somatiri, N.L.D.; Wu, M. Electrochemical characteristics of polyaniline cathodes and anodes in aqueous electrolytes. *Mol. Cryst. Liq. Cryst.* 121:187–190.

51. Chen, S.A.; Fang, Y. 1993. Polyaniline Schottky barrier: Effect of doping on rectification and photovoltaic characteristics. *Synth. Met.* 60:215–222.

52. Herod, T.E.; Schlenoff, J.B. 1993. Doping-induced strain in polyaniline: Stretchoelectrochemistry. *Chem. Mater.* 5:951–955.

53. Kaneto, K.; Kaneko, M.; Min, Y.; MacDiarmid, A.G. 1995. Artificial muscle: Electromechanical actuators using polyaniline films. *Synth. Met.* 71:2211–2212.

54. Kapli, A.; Taunk, M.; Chand, S. 2009. Preparation and characterization of chemically synthesized poly(N-methylaniline). *Synth. Met.* 159:1267–1271.

55. Yang, M.; Xiang, Z.; Wang, G. 2012. A novel orchid-like polyaniline superstructure by solvent–thermal method. *J. Colloid Interface Sci.* 367:49–54.

56. Lallemand, F.; Hihn, J.-Y.; Atobe, M.; Taouil, A.E. 2012. Sonoelectropolymerization, In Pollet B. (Ed.), *Power Ultrasound in Electrochemistry: From Versatile Laboratory Tool to Engineering Solution*, 249–276. John Wiley & Sons, London.

57. Osawa, S.; Ito, M.; Tanaka, K.; Kuwano, J. 1987. Electrochemical polymerization of thiophene under ultrasonic field. *Synth. Met.* 18:145–150.

58. Atobe, M.; Fuwa, S.; Sato, N.; Nonaka, T. 1997. Ultrasonic effects on electroorganic processes. Part 5. Preparation of a high density polyaniline film by electrooxidative polymerization under sonication. *Denki Kagaku* 65:495–497.

59. Atobe, M.; Nonaka, T. 1998. New developments in sonoelectrochemistry. *Nippon Kagaku Kaishi*, 219–230.

60. Atobe, M.; Kaburagi, T.; Nonaka, T. 1999. Ultrasonic effects on electroorganic processes. Part 13. A role of ultrasonic cavitation in electrooxidative polymerization of aniline. *Electrochemistry* 67:1114–1116.

61. Atobe, M.; Tsuji, H.; Asami, R.; Fuchigami, T. 2006. A study on doping-undoping properties of polypyrrole films electropolymerized under ultrasonication. *J. Electrochem. Soc.* 153:D10–D13.

62. Taouil, A.E.; Lallemand, F.; Hihn, J.Y.; Blondeau-Patissier, V. 2011. Electrosynthesis and characterization of conducting polypyrrole elaborated under high frequency ultrasound irradiation. *Ultrason. Sonochem.* 18:907–910.

63. Taouil, A.E.; Lallemand, F.; Hihn, J.Y.; Melot, J.M.; Blondeau-Patissier, V.; Lakard, B. 2011. Doping properties of PEDOT films electrosynthesized under high frequency ultrasound irradiation. *Ultrason. Sonochem.* 18:140–148.

64. Walton, D.J.; Iniesta, J.; Plattes, M.; Mason, T.J.; Lorimer, J.P.; Ryley, S.; Phull, S.S.; Chyla, A.; Heptinstall, J.; Thiemann, T.; Fuji, H.; Mataka, S.; Tanaka, Y. 2003. Sonoelectrochemical effects in electro-organic systems. *Ultrason. Sonochem.* 10:209–216.

5 Novel Electrode Materials in Electroorganic Synthesis

Christoph Gütz, Sebastian Herold,
and Siegfried R. Waldvogel
Johannes Gutenberg-Universität Mainz

CONTENTS

5.1 Introduction to Boron-Doped Diamond Electrodes 97
5.2 Cathodic Conversions at Boron-Doped Diamond Electrodes 99
    5.2.1 Carbon Dioxide Reduction 99
    5.2.2 Cathodic Reductive Coupling of Methyl Cinnamates 100
    5.2.3 Reduction of Oximes 101
5.3 Anodic Transformations at Boron-Doped Diamond Electrodes 103
    5.3.1 Alkoxylation Reactions...103
    5.3.2 Fluorination 104
    5.3.3 Cyanation..104
    5.3.4 Cleavage of C,C-Bonds..105
    5.3.5 Electrochemical Amination Reactions..105
    5.3.6 Anodic Phenol Coupling Reaction 107
5.4 Novel Concepts and Electrode Materials for Electrochemical Reductions.......110
    5.4.1 Improvement of Existing Cathodes 110
    5.4.2 Novel Cathode Materials 111
5.5 Summary 119
References...119

5.1 INTRODUCTION TO BORON-DOPED DIAMOND ELECTRODES

Currently, polycrystalline boron-doped diamond (BDD) films are mostly prepared by two methods on various substrates, e.g., silicon, niobium, titanium, and different carbon-based materials. Either a microwave plasma is employed using acetone and trimethyl borate as the starting material (microwave plasma-enhanced chemical vapor deposition) or thermal decomposition techniques (e.g., hot-filament chemical vapor deposition, HFCVD) are applied [1–3]. In the latter case, the carbon source is usually methane and diborane serves as the boron source. Only HFCVD is currently commercially exploited; both, large electrode geometries (>0.5×0.5 m) and highly sophisticated electrode designs are accessible by this method. With HFCVD methods, diamond films of up to 50 μm thickness can be obtained. Since methane can originate from renewable resources, BDD electrodes are usually considered as sustainable electrode material. Doping diamond with boron leads to a semi-conducting material with a narrow bandgap. Usually, the boron content is adjusted within the range of 200–5000 ppm. However, for the application in organic media, the diamond coating has to be pinhole-free since corrosion of the support material will cause a collapse of the diamond film [4]. In contrast, aqueous media promote the formation of insulating and stable oxides leading to inactive but still stable electrode sections [2]. Because of the strong σ-bonds of the sp^3-hybridized carbon atoms, diamond exhibits an exceptional chemical and electrochemical stability. Outstanding properties of BDD are the unusually large offset potentials for the evolution of hydrogen (−1.1 V) and the evolution of oxygen (2.3 V) in aqueous media (Figure 5.1a). BDD shows an even larger electrochemical window in fluorinated alcohols like 1,1,1,3,3,3-hexafluoroisopropanol with an anodic limit of 3.2 V and a cathodic limit of almost −2 V (Figure 5.1b). This system has an accessible potential range of roughly 5 V which is still the largest for a protic electrolyte [5]. A limitation might be caused by the supporting electrolytes employed [6]. In the example depicted, methyltriethylammonium methylsulfate was used.

Hydroxyl radicals and ozone can be formed in aqueous solution at highly positive potentials. The formation of OH radicals was shown using spin traps [7]. Furthermore, anodic formation of methoxy radicals in a solution containing methanol has been proposed [8,9]. Since hydroxyl radicals exhibit an utmost reactivity, degradation processes up to complete mineralization [10,11] were studied and exploited for wastewater treatment [1,12]. Furthermore, BDD has been successfully employed in the industrial synthesis of inorganic chemicals, e.g., persulfates [13]. Besides the aforementioned destructive performance of BDD electrodes, their extraordinary characteristics, e.g., formation of OH spin-centers, have shown great success in electrosynthetic transformations [14]. BDD exhibits two different modes of action: first, at highly positive potentials and low concentrations of substrate, a complete degradation of organic compounds to CO_2 and H_2O takes place. These are usually the conditions applied for wastewater treatment. Second, at intermediate potentials, a direct electron transfer can occur which makes electrosynthetic transformations

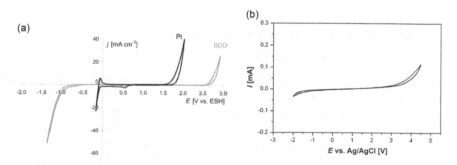

FIGURE 5.1   (a) CV of BDD and Pt in H_2O. (b) CV of BDD in $(CF_3)_2CHOH$.

possible. Therefore, a low current density combined with high concentrations of substrate leads to partial oxidation which can be exploited to generate complex organic compounds [15]. The challenge for organic synthesis is to efficiently circumvent the first pathway in order to use this innovative electrode material for constructive organic synthesis. Furthermore, a stabilization of reactive intermediates is beneficial. Especially, fluorinated alcohols like hexafluoroisopropanol (HFIP) can enhance the lifetime of spin-centers dramatically [16–18]. A particular solvent effect of these solvents with an extraordinary hydrogen bonding capability is proposed [19,20].

5.2   CATHODIC CONVERSIONS AT BORON-DOPED DIAMOND ELECTRODES

In contrast to anodic transformations, electrochemical reduction at BDD is still less studied. Usually, cathode materials should exhibit a high overpotential for the evolution of hydrogen. This is necessary as electrochemical reductions are usually carried out in protic media. Heavy metals like lead, mercury, and cadmium have been established and usually show good to excellent results. However, a substitution of these toxic heavy metals could have an enormous academic as well as technical impact. In this particular area of electroorganic synthesis, BDD seems to be the ideal material because of the large offset potential for hydrogen evolution which is comparable to lead or mercury and the outstanding electrochemical stability [12]. It is noteworthy that during electrosynthesis, many heavy metal electrodes show cathodic corrosion leading to undesired contamination with toxic metals [21–23]. Cathodic dissolution of noble metals, e.g., platinum, is observed at very negative potentials [24].

5.2.1   CARBON DIOXIDE REDUCTION

Carbon dioxide (CO_2) is an abundant and highly attractive C1 building block. Because of this, and also due to ecological considerations, cathodic activation of CO_2 and subsequent incorporation into high-value chemical products is of upcoming interest [25]. BDD was used for the reductive carboxylation of the aldehyde 1 to 2-hydroxy-4-methylsulfanylbutyric acid (2) which represents an important technical product used for animal feeding. In the electrochemical approach, an Mg sacrificial anode and a BDD cathode can be used (conversion 66% and current efficiency, CE, 22%). However, one drawback of this reaction is the direct reduction of aldehyde 1 to alcohol 3 (Figure 5.2) [26].

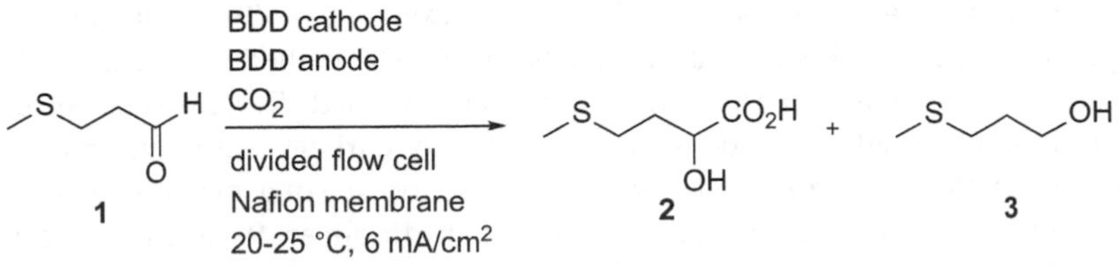

FIGURE 5.2   Reductive carboxylation of 3-(methylthio)propionaldehyde (1) to 2-hydroxy-4-methylsulfanylbutyric acid (2).

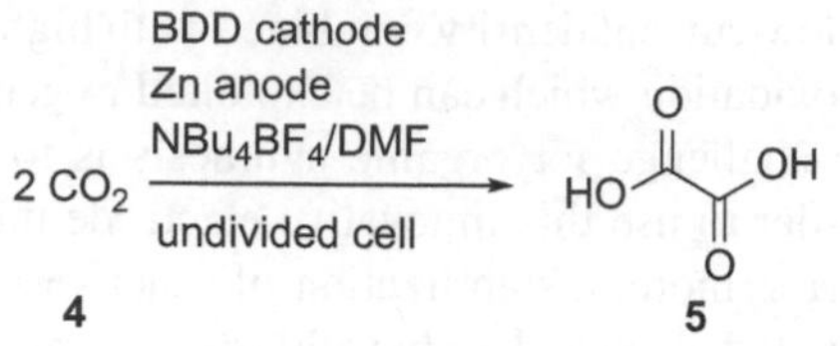

$$2\ CO_2 \xrightarrow[\text{undivided cell}]{\substack{\text{BDD cathode} \\ \text{Zn anode} \\ \text{NBu}_4\text{BF}_4/\text{DMF}}} \text{(oxalic acid)}$$

4   5

FIGURE 5.3   Hydrodimerization of carbon dioxide to oxalic acid (**5**).

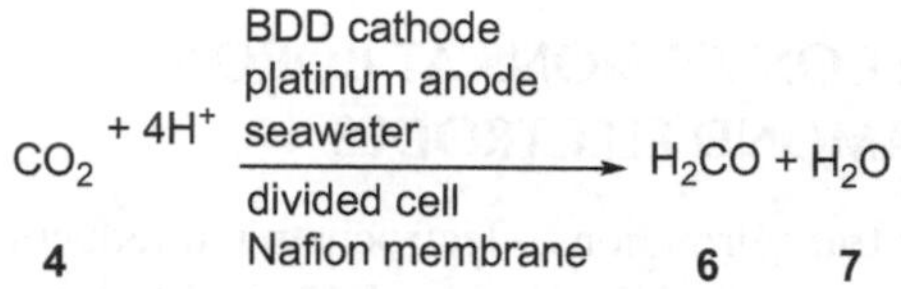

$$CO_2 + 4H^+ \xrightarrow[\substack{\text{divided cell} \\ \text{Nafion membrane}}]{\substack{\text{BDD cathode} \\ \text{platinum anode} \\ \text{seawater}}} H_2CO + H_2O$$

4   6   7

FIGURE 5.4   Electrochemical generation of formaldehyde (**6**) from carbon dioxide on BDD cathodes.

Furthermore, it seems to be possible that a direct hydrodimerization of carbon dioxide to obtain oxalic acid (**5**) is possible. For this transformation, an undivided cell and zinc as sacrificial anode material is used. The electrolysis is conducted in dimethylformamide (DMF) with 12 mM $NBu_4^+ BF_4^-$ at a current density of 6 mA cm^{-2}. With this setup, a CE of 60% is accomplished (Figure 5.3) [27].

In 2014, Einaga and Nakata reported the highly selective electrochemical generation of formaldehyde (**6**) from carbon dioxide using BDD electrodes. Formaldehyde is an important bulk chemical, which is used on industrial scale for the production of polymers. The maximum Faradaic efficiency for formaldehyde formation was 74% in an MeOH electrolyte. Formic acid was obtained as a by-product. Other metal cathodes, e.g., Cu, Sn, Ag, and W, showed no performance at all for the formation of formaldehyde. Glassy carbon as anode material yielded formaldehyde in 19% Faradaic yield. Most remarkably, the transformation can also be conducted in seawater as the electrolyte. However, the Faradaic efficiency is lower than in an MeOH-based electrolyte (Figure 5.4) [28].

5.2.2   CATHODIC REDUCTIVE COUPLING OF METHYL CINNAMATES

Lignans and neolignans are widely found as secondary plant metabolites and have an interesting biological profile. The cathodic coupling of methyl cinnamate (**8a**) at BDD electrodes leads to dimethyl 3,4-diphenylhexanedioate (**9**) which can be used to synthesize new neolignan-type products. At a BDD cathode, a yield of 85% of hydrodimer **9** (racemate/meso=74:26) could be obtained. Several other electrode materials were tested as cathodes, including glassy carbon, platinum, and magnesium. When Pt and Mg cathodes were applied, the starting material was almost completely recovered. Glassy carbon yielded hydrodimer **9** in 34% (racemate/meso=74:26) and cyclization product **10** in 25% along with 41% of **8b**. Metallic cathodes, like Hg, Cu, Pb, Zn, Sn, or Ag, yielded as major product the five-membered ring **10** by a subsequent Dieckmann condensation. BDD was superior to all these tested

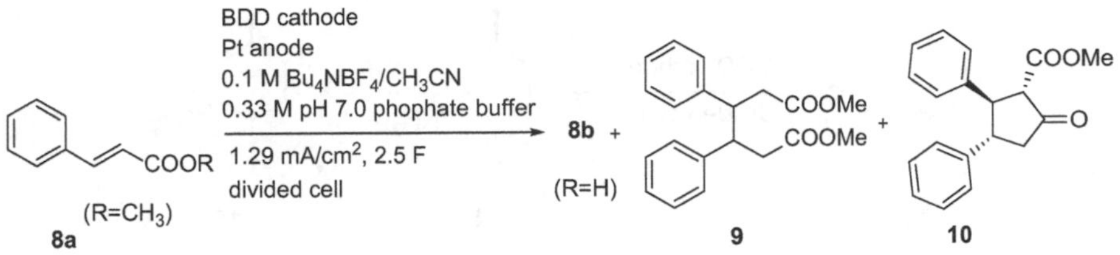

FIGURE 5.5 Cathodic reductive coupling of methyl cinnamate (**8a**).

cathode materials in respect to the selectivity in formation of the desired hydrodimer **9** (Figure 5.5) [29].

5.2.3 REDUCTION OF OXIMES

A common strategy for the synthesis of amines is the reduction of the corresponding oximes. Conventionally, this reduction is carried out under Bouveault-Blanc-type reaction conditions with alkali metals in the presence of a proton source [30]. However, this kind of reduction can also be carried out using electrochemical methods. This is beneficial in terms of safety aspects since large amounts of alkali metals can be circumvented. In addition, it is much more ecological because the production of reagent waste can be avoided. Mostly, reductions at a mercury pool are described, since the high overpotential for hydrogen evolution at this electrode material in protic solvents proves to be suitable [31]. A BDD cathode was successfully employed in the reduction of cyclopropyl phenylketone oxime (**11**) to racemic α-cyclopropyl benzylamine (**12**) at anhydrous conditions (Figure 5.6). Control experiments with Pb cathode gave similar results at the same electrolysis conditions [32]. This electrochemical methodology is superior to catalytic hydrogenation wherein significant amounts of ring opening occur, leading to tedious purification issues.

These initial studies lead to investigations of the more congested menthone oxime **11** as substrate. Menthone oxime **13** could yield both diastereomers, (+)-neomenthylamine (**14**) and (−)-menthylamine (**15**) upon reduction (Figure 5.7). Both diastereomers are unique in their properties and applications [33–41]. Studies with BDD cathodes revealed that **13** seems to be a non-suitable substrate, as almost no conversion is detected. This indicates that sterical issues seem to play a crucial role at BBD cathodes, and the electroconversion takes place directly on the electrode surface. Cathodic treatment using lead results in an almost quantitative reduction [42].

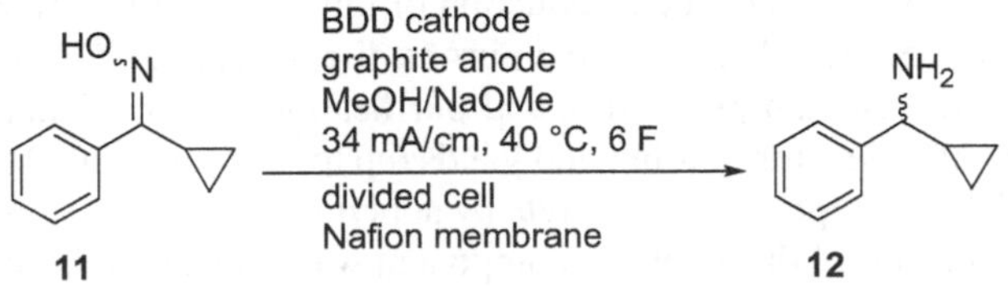

FIGURE 5.6 Reduction of cyclopropyl phenylketone oxime (**11**) at a BDD cathode.

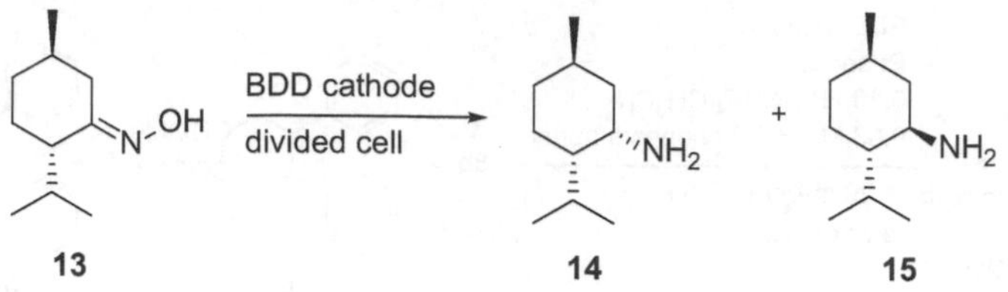

FIGURE 5.7 Reduction of menthone oxime (**13**) at a BDD cathode.

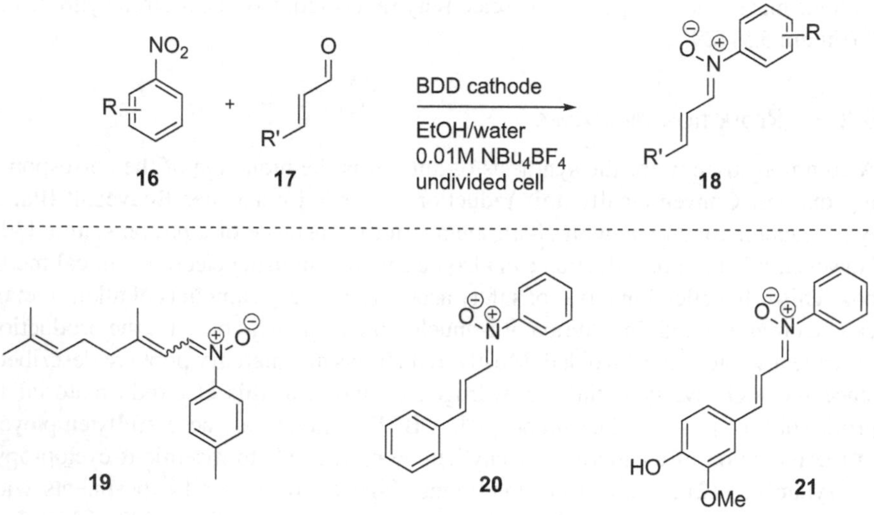

FIGURE 5.8 Reduction of nitroarenes (**16**) in the presence of aldehydes (**17**) to nitrones (**18**).

In the reduction of nitroarenes, BDD cathodes seem to be a superior material and can be used for the direct synthesis of nitrones **18** [43]. Several synthetic approaches to nitrones **18** have been established. However, these protocols mostly require multiple synthetic steps and sacrificial metals, and a precise control of the reaction conditions to avoid mixtures of products. In this context, the use of an electroreduction approach with BDD electrodes for the synthesis of nitrones was developed by using green solvent conditions with aldehydes **17** and nitro compounds **16** as the starting materials. It is well known that cathodic reduction of aldehydes easily generates alcohols or pinacols, but the electroreduction of nitro derivatives can be attained in the presence of the aldehyde to yield the nitrone directly (Figure 5.8), reducing operational steps. Downstream processing and purification are significantly facilitated. All enals used afforded nitrones in good yields (up to 72%) by using BDD cathode in 0.01 M NBu_4BF_4 as supporting electrolyte by applying 7.5 mA cm^{-2}. Interestingly, no formation of pinacols of the aldehyde component was detected. It is noteworthy that BDD provides not only the best yield for nitrones **18** but also the cleanest and highest conversion. In addition, this protocol turned out to be scalable as well.

5.3 ANODIC TRANSFORMATIONS AT BORON-DOPED DIAMOND ELECTRODES

5.3.1 ALKOXYLATION REACTIONS

Direct anodic C,H activation of hydrocarbons for the installation of an alkoxy moiety is of tremendous importance and a contemporary area of investigation. As the most intriguing features of BDD electrodes are the efficient generation of hydroxyl and alkoxyl radicals, these electrodes seem to be perfect for anodic alkoxylation reactions. Electrogenerated alkoxy radicals may be implemented directly into the product or serve as a strong oxidant. Trimethylorthoformate (**23**) represents an attractive C1 building block as well as a formic acid equivalent. Therefore, an interesting access to **23** is an anodic alkoxylation of formaldehyde dimethylacetal (**22**). This electrochemical reaction can be carried out at BDD anodes (Figure 5.9) [44].

Substituted aromatic aldehydes are important bulk chemicals for the production of commercial products. A typical example is *p-tert*-butylbenzaldehyde which is produced on a multiton scale. This aromatic aldehyde can be obtained by electroorganic oxidation of *p-tert*-butyltoluene (**24**) to *p-tert*-butylbenzaldehyde dimethylketal (**25**) in methanol followed by hydrolysis. When BDD is applied as anode, an accumulation of a dibenzylic species is observed. However, in the course of the electrolysis, the anodic cleavage of this intermediate and dimethoxylation provides the desired dimethylketal **25** (Figure 5.10) [8,45].

The anodic conversion of furan (**26**) in methanol yields 2,5-dimethoxy-2,5-dihydrofuran (**27**) which is a twofold protected 1,4-dialdehyde and is commonly used as the C4 building block in heterocyclic synthesis (Figure 5.11). The electrolysis is usually carried out by using a bromide-mediated system in combination with a

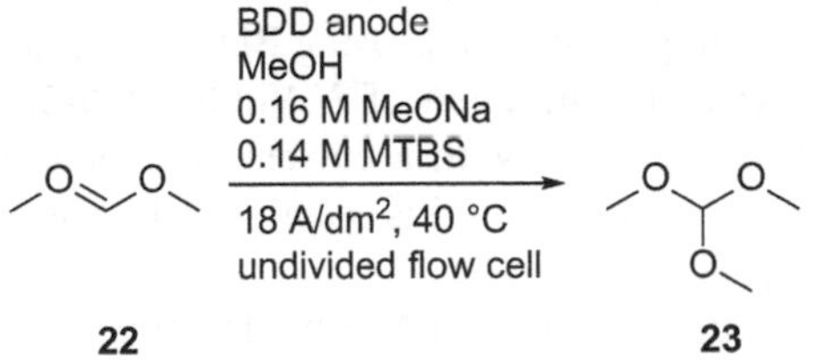

FIGURE 5.9 Anodic alkoxylation of formaldehyde dimethylacetal (**22**) at BDD electrodes.

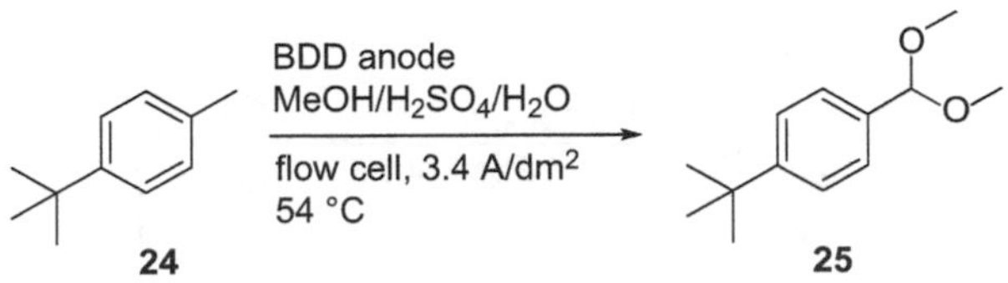

FIGURE 5.10 Electrooxidative synthesis of the important bulk chemical *p-tert*-butylbenzaldehyde dimethylketal (**25**) using BDD electrodes.

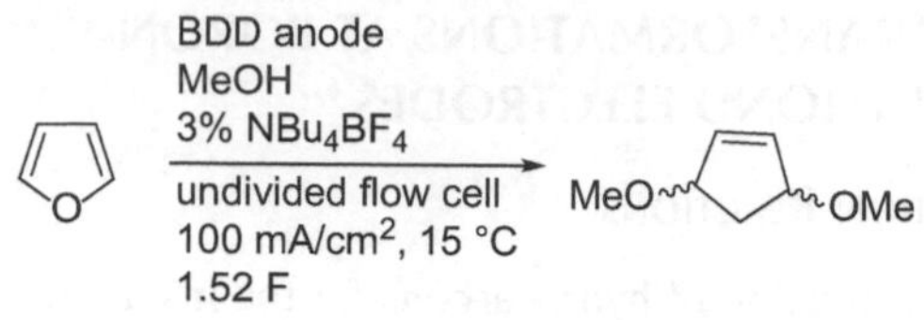

FIGURE 5.11 Mediator-free dimethoxylation of furan (**26**) at BDD anodes.

platinum anode [46,47]. However, this transformation can be conducted with BDD anodes wherein no mediator is required [48].

5.3.2 Fluorination

Anodic fluorinations are well known, and the chemical as well as electrochemical stability of BDD makes this material perfect for fluorination reactions [49]. Usually, α to heteroatom-substituted position are accessible by this method, which is performed using hydrogen fluoride/trimethylamine mixtures as the fluoride source. Different electrode materials were studied by Fuchigami and co-workers for the fluorination of oxindole **28**. BDD anodes provide comparable yields like platinum, whereas glassy carbon proved to be less efficient (Figure 5.12) [50,51].

Furthermore, the anodic fluorination of 1,4-difluorobenzene was reported using a BDD anode yielding, on the analytical scale, tetrafluoro cyclohexadiene [52].

5.3.3 Cyanation

Similarly, electrochemical cyanation reactions provide access to nitriles without previous introduction of leaving groups [53]. Using BDD as the anode material, a cyano moiety was installed α to the nitrogen of N,N-dibutyl-N-2,2,2-trifluorethylamine (**30**) yielding the amino nitrile **31** with a current efficiency of 77% (Figure 5.13). Platinum provides the same results in this reaction, whereas glassy carbon showed no performance at all [54].

Furthermore, the anodic cyanation of N-methylpyrrole (**32**) was performed at the same conditions. Mono **33** and dicyanated **34** products were obtained in this case (Figure 5.14) [54].

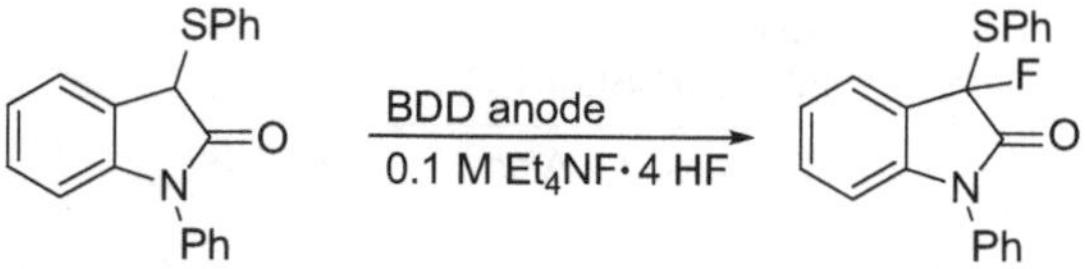

FIGURE 5.12 Anodic fluorination of oxindole (**28**) using BDD electrodes.

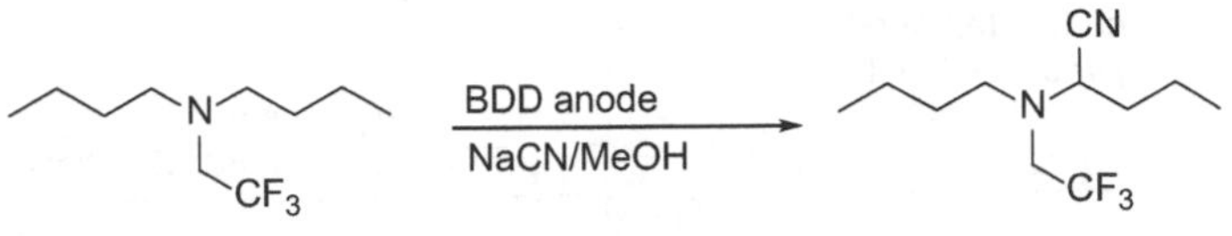

FIGURE 5.13   Anodic cyanation of the amine (**30**) leading to the amino nitrile (**31**).

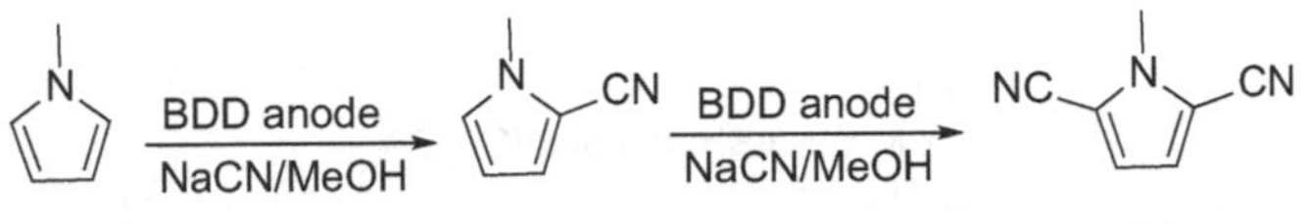

FIGURE 5.14   Anodic conversion of pyrrole (**32**) to the mono- and dicyanated products (**33**) and (**34**).

5.3.4   CLEAVAGE OF C,C-BONDS

Schäfer and co-workers demonstrated that aliphatic olefins proved to be stable toward oxidation with BDD [55]. However, more activated olefins such as phenanthrene can successfully be cleaved yielding the desired product **36** in 39% yield (by gas chromatography) as well as 2,2′-bis-dimethoxymethylbiphenyl (**37**) in 15% yield (by gas chromatography) (Figure 5.15) [32].

5.3.5   ELECTROCHEMICAL AMINATION REACTIONS

Aniline derivatives are important building blocks in organic synthesis [56–60]. An elegant electrosynthetic approach to such structural motifs was introduced by Yoshida and co-workers in 2013 [61,62]. The presented methodology allows a mild and regioselective C,H amination of activated arenes **38** via anodic pyridination. Subsequently, ring opening of the obtained Zincke intermediates **39** yield the desired anilines **40** (Figure 5.16).

Waldvogel and co-workers effectively applied this cutting-edge amination technology to less-activated arenes, e.g., simple alkylated aromatic compounds [63].

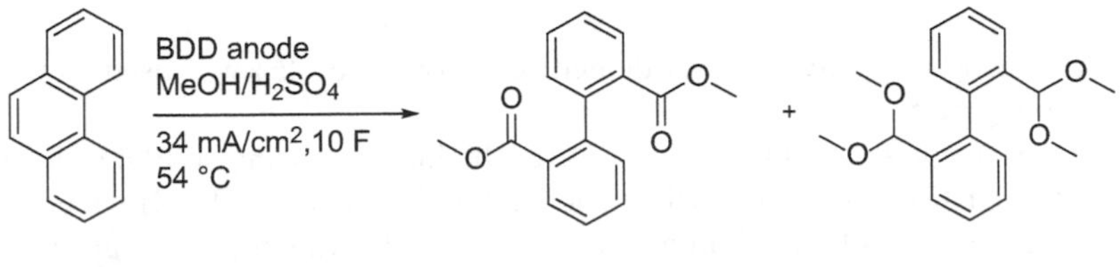

FIGURE 5.15   Anodic cleavage of activated C,C-double bonds on BDD anodes.

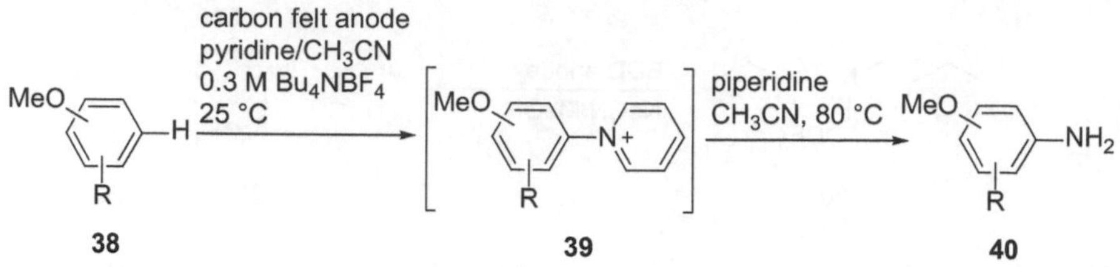

FIGURE 5.16   Anodic amination of electron-rich arenes.

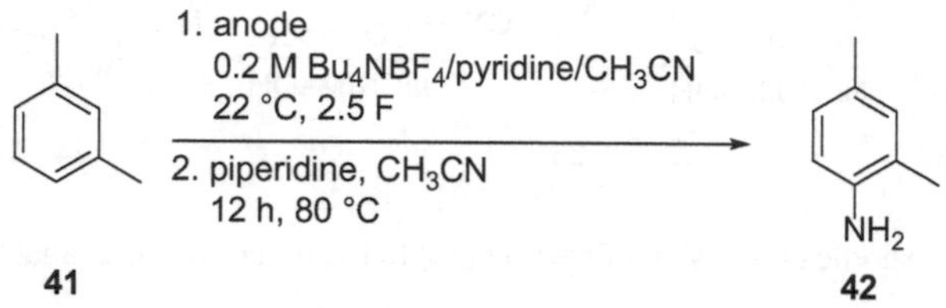

FIGURE 5.17   Anodic amination of *m*-xylene (**41**) employed as test substrate for screening reactions with different anode materials.

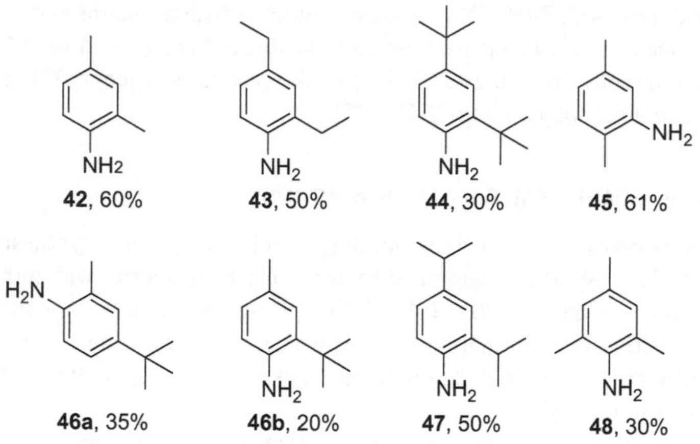

FIGURE 5.18   Substrate scope of the electrochemical amination of simple alkylated arenes using BDD anodes.

Different anode materials were evaluated, and *m*-xylene (**35**) was used as the test substrate (Figure 5.17).

Best results of up to 60% of **42** were obtained by using a BDD electrode. Optimized reaction conditions (BDD anode, 2.5 F, 8 mA/cm^2, 22°C) were applied to a broad scope of alkylated arenes. A 1,3-alkyl substitution pattern gives rise to aniline derivatives in yields of up to 60%. Even *tert*-butyl moieties are compatible with this method despite cationic intermediates. Usually, the sterically less hindered product is formed (Figure 5.18).

5.3.6 ANODIC PHENOL COUPLING REACTION

Biphenols are important structural motifs not only in natural products [64] but also for technical applications [65]. In this regard, 3,3′,5,5′-tetramethyl-2,2′-biphenol (50) has attracted significant attention. By an electrochemical approach, costly and toxic transition metals as well as the use of stoichiometric amount of oxidants for the synthesis of such biaryls can be avoided [66]. However, a direct electrochemical coupling of 2,4-dimethylphenol (49) is challenging, as a number of by-products like Pummerer's ketone 51 and pentacyclic scaffolds, e.g., 52, are formed in significant amounts (Figure 5.19) [15,67,68].

Surprisingly, this challenge was tackled by the use of BDD anodes. Almost neat phenol 49 with 11% water and an ammonium salt as supporting electrolyte yielded the desired *ortho,ortho*-coupling products 50 in good selectivity (up to 19:1, 50/51) and yields of up to 49% upon oxidation at BDD electrodes [15]. In this case, the high concentration of substrate 49 compensates for the high reactivity of intermediate oxyl radicals, leading to the electrosynthetic success. However, other phenols show no detectable products or were mineralized. To overcome this limitation, fluorinated alcohols, e.g., 2,2,2-trifluorethanol and HFIP, were used in the electrolyte (Figure 5.20) [69].

Using the elaborated protocol with HFIP in combination with a BDD anode, non-symmetrical (*ortho,meta*) coupling products can be isolated [70]. A broad range of guaiacol derivatives were studied and, depending on the steric demand of the

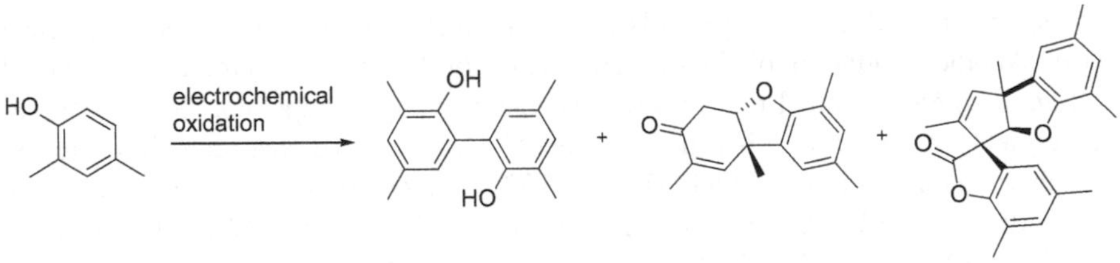

FIGURE 5.19 Anodic treatment of 2,4-dimethylphenol (49).

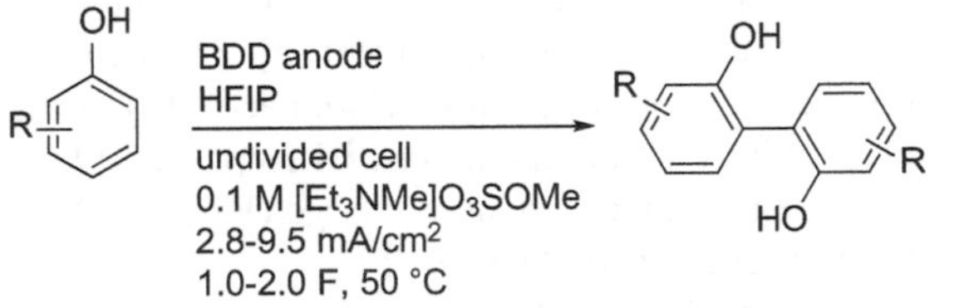

FIGURE 5.20 Anodic homocoupling of phenols at BDD electrodes using fluorinated alcohols as solvents.

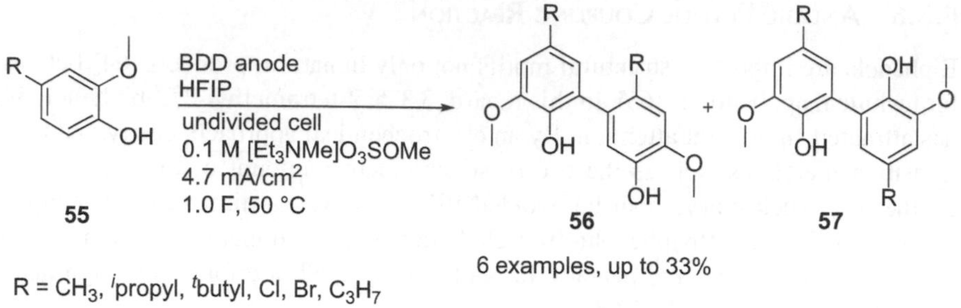
BDD anode
HFIP
undivided cell
0.1 M [Et3NMe]O3SOMe
4.7 mA/cm2
1.0 F, 50 °C
55
56
57
6 examples, up to 33%
R = CH3, ipropyl, tbutyl, Cl, Br, C3H7

FIGURE 5.21 Non-symmetrical (**56**) or symmetrical (**57**) biphenols accessible via anodic homocoupling of phenols at BDD electrodes.

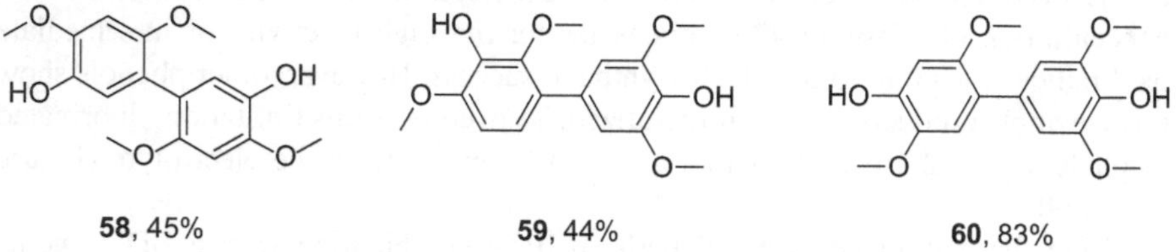
58, 45%
59, 44%
60, 83%

FIGURE 5.22 Homocoupling products of electron-rich dimethoxy phenols.

substituent in position 4 of **55**, either symmetrical **57** or non-symmetrical biphenols **56** were obtained (Figure 5.21).

Electron-rich dimethoxy phenols give selectively one coupling product in each case. Anodic oxidation of 2,4-dimethoxyphenol leads to a symmetric biphenol (*meta,meta*) **58**. Syringol provides the *para,meta*-coupled biphenol **59** in a comparable yield. In contrast, 2,5-dimethoxyphenol forms a *para,para*-dehydrodimer **60** which was isolated in good yield (Figure 5.22). The reactions were conducted in an undivided cell with a BDD anode at a current density of 4.7 mA/cm^2 and 50°C. A 0.1 M $[Et_3NMe]O_3SOMe$/HFIP solution was used as the electrolyte [70].

One of the key features of oxidative cross-coupling reactions is a selective oxidation of one component which then exclusively attacks the other component. As a *ortho,meta*-coupled phenol was observed in the anodic conversion of 4-methylguaiacol, it seemed possible that such a coupling should also happen with a phenol and another arene component. And indeed anodic phenol-arene cross-coupling is viable using a BDD anode in combination with HFIP as solvent [5]. A broad range of arenes **62** were transformed in this reaction with 4-methyl guaiacol (**61**) as the phenol component. The obtained selectivity was in the range of 1:1 up to >50:1 (**63**:**64**) depending on the applied current density and amount of charge (Figure 5.23). Surprisingly, no phenol homocoupling products were detected [5].

A more detailed study of this novel anodic cross-coupling reaction revealed that the addition of 9% H_2O or 18% MeOH as protic additives to the electrolyte is beneficial for the transformation. Yields of up to 70% and selectivity of up >100:1 for the cross-coupling products were obtained [71]. The additive allows a shift in oxidation potential, triggering the selectivity for the cross-coupling reaction [19].

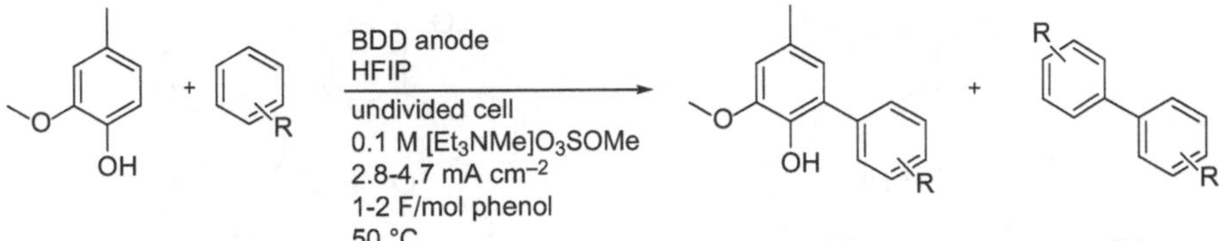

61 62 63 64

R = MeO, Br, CH_3

8 examples, 11-47%

FIGURE 5.23 Anodic phenol-arene cross-coupling at BDD electrodes.

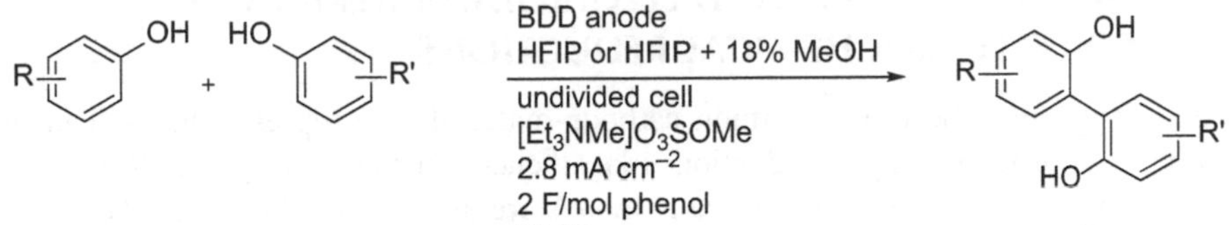

65 66 67

R' = CH_3, ^{i}propyl, ^{t}butyl, MeO, Br

14 examples, 7-63%

FIGURE 5.24 Anodic phenol–phenol cross-coupling at BDD electrodes.

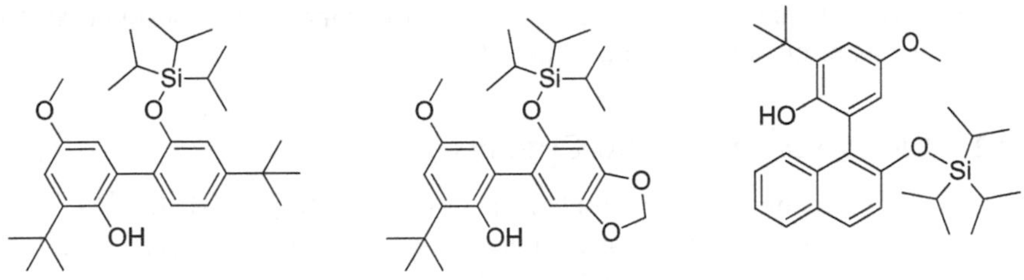

68, 92% **69**, 82% **70**, 81%

FIGURE 5.25 Partially protected biphenols via anodic cross-coupling.

Recently, this anodic coupling method was expanded to phenol–phenol cross-coupling [72]. In almost all examples, no homocoupling of the individual phenolic components was observed. As phenol **65** either 4-methyl guaiacol or 2-naphthol were used. These phenols exhibit a lower oxidation potential than phenols **66** thus initiating the reaction sequence (Figure 5.24).

The introduction of suitable silyl protective groups gave rise to a new robust and scalable conversion at BDD anodes (Figure 5.25) [73].

The approach can be employed to perform this conversion two times in order to obtain symmetric and non-symmetric *m*-terphenyl-2,2″-diols (Figure 5.26) [74].

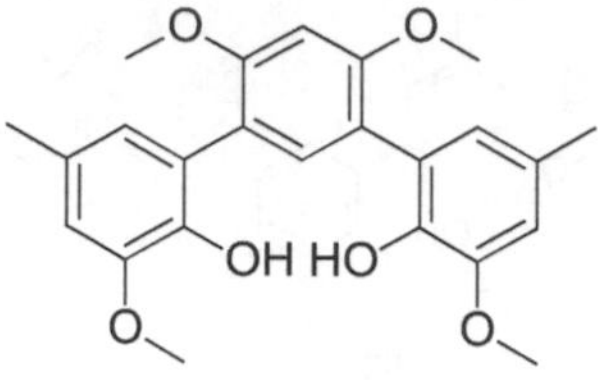

71, 38% 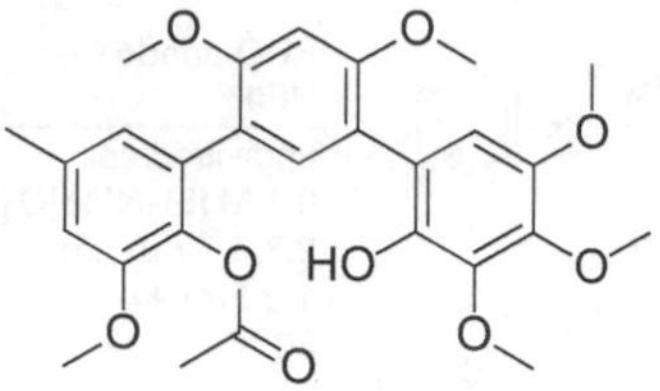   **72**, 84%

FIGURE 5.26 Symmetric and non-symmetric m-terphenyl-2,2'-diols.

5.4 NOVEL CONCEPTS AND ELECTRODE MATERIALS FOR ELECTROCHEMICAL REDUCTIONS

For a long time, the most common cathode materials are represented by lead or mercury. A broad variety of reductions like alkene–alkene couplings [75–77], dehalogenation reactions [78–82], and many more were described in literature. The most important advantage of these materials is the high overpotential for hydrogen evolution in protic media. However, their chemical stability during electrolysis is highly challenging. Impurities of lead or mercury can be found in the latter product—even after strict purification protocols [83–87]. These impurities are highly problematic with regard to the synthesis of fine chemicals or pharmaceutically relevant compounds due to their toxicity. Other well-established materials like glassy carbon or platinum are only suitable at a significant less-reductive regime or aprotic media. Nevertheless, there are only few investigations on the improvement of standard electrode materials or on novel cathode materials.

5.4.1 IMPROVEMENT OF EXISTING CATHODES

A first step to circumvent these problems is to modify already existing electrodes, since it is easier then establishing complete novel ones. The concept of modified electrodes is well known in the field of bioelectrochemistry and electrocatalysis for defined transformations, but not in electrosynthesis for a broad range of reactions. Initially, this concept was used by our group for the application of lead electrodes in electroorganic synthesis [23,42].

The key for a higher stability of lead cathodes was the application of polyammonium salts as an additive: The polyammonium salt forms an ionic coating at the cathode during electrolysis. This has several advantages. First, the decoration at the cathode successfully suppresses the corrosion of the electrode [23]. Second, hydrogen evolution, as a side reaction, was prohibited more effectively. The coat inhibits the diffusion of positively charged protons to the electrode (Figure 5.26a). This effect could also be detected by cyclic voltammetry (CV) measurements, wherein a higher overpotential of $-157\,mV$ for the hydrogen evolution was observed (in a methanol solution with 2% H_2SO_4, Figure 5.27b). Thereby, a quantitative yield was achieved and the CE of the conversion was increased dramatically [23].

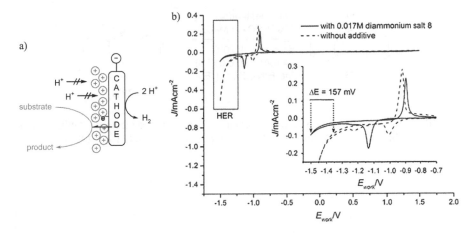

FIGURE 5.27 (a) Schematic concept of the polyammonium salt functionalized electrode surface (left side) and without the additive (right side). (b) CV of a decorated and an undecorated lead electrode in methanol+2% H_2SO_4 (HER=hydrogen evolution reaction). (With permission from Edinger, C. et al., *ChemElectroChem.*, 1, 1018, 2014.)

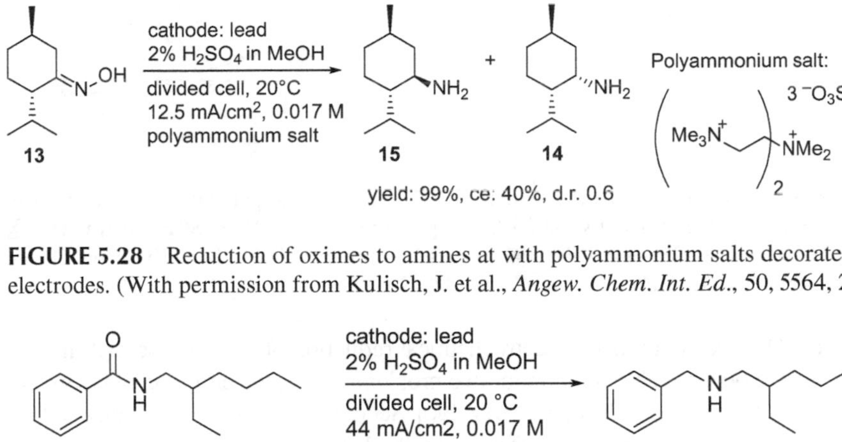

FIGURE 5.28 Reduction of oximes to amines at with polyammonium salts decorated lead electrodes. (With permission from Kulisch, J. et al., *Angew. Chem. Int. Ed.*, 50, 5564, 2011.)

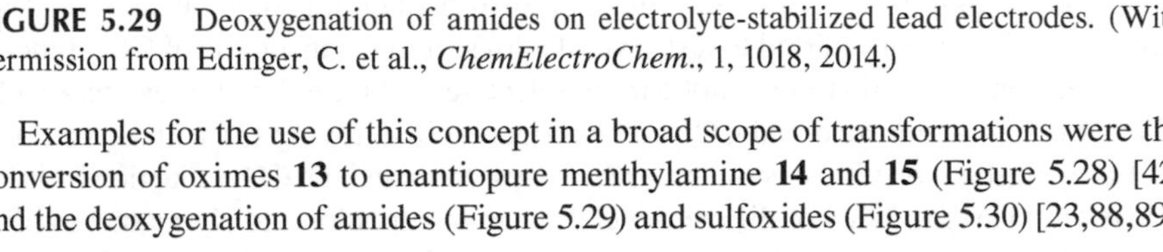

FIGURE 5.29 Deoxygenation of amides on electrolyte-stabilized lead electrodes. (With permission from Edinger, C. et al., *ChemElectroChem.*, 1, 1018, 2014.)

Examples for the use of this concept in a broad scope of transformations were the conversion of oximes **13** to enantiopure menthylamine **14** and **15** (Figure 5.28) [42] and the deoxygenation of amides (Figure 5.29) and sulfoxides (Figure 5.30) [23,88,89].

5.4.2 Novel Cathode Materials

Beside the improvements for existing electrode materials, the development of novel cathode materials for electroorganic synthesis is even more challenging and rarely

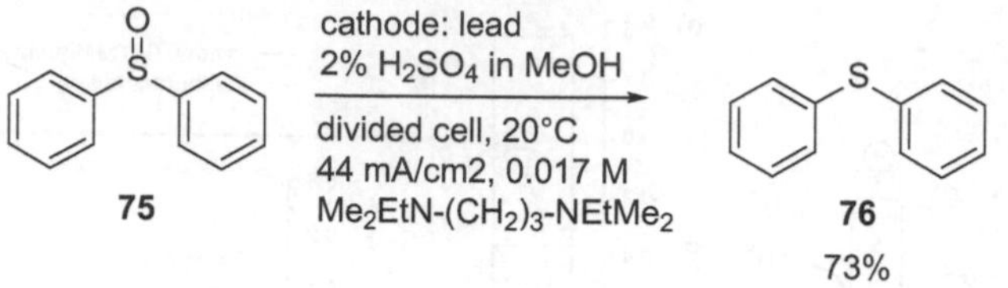

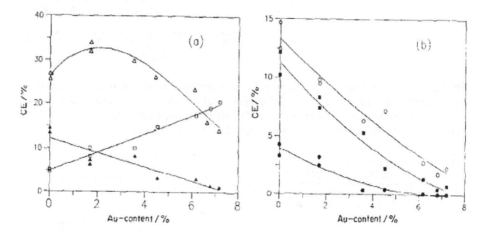

FIGURE 5.30 Deoxygenation of sulfoxides on a decorated lead electrode. (With permission from Edinger, C. and Waldvogel, S.R., *Eur. J. Org. Chem.*, 2014, 5144, 2014.)

FIGURE 5.31 Mean CE for the products of CO_2 reduction vs. percentage gold content of the copper electrode in 0.1 M KCl at 25°C at a potential of −1.85 V vs. SCE: (a) Δ C_2H_4, ▲ CH_4, ☐ CO; (b) ○ EtOH, ● *n*-PrOH, ■ HCOOH. (With permission from Kyriacou, G. and Anagnostopoulos, A., *J. Electroanal. Chem.*, 328, 233, 1992.)

discussed. One exception is the electrochemical reduction of CO_2. These high-impact results tend to be a possible solution for two problems. First, the emission of the greenhouse gas CO_2 has to be reduced stepwise [90]. Second, because the avoidance of CO_2 emission is only partially possible, effective methods for the conversion of CO_2 to valuable C1 or C2 building blocks like methanol, formate, or ethanol have to be invented.

In 1991, Watanabe et al. demonstrated that with pure metal electrodes like Cu, Sn, Pt, or Pb, no high faradic efficiency and no low overpotential could be achieved [91]. Therefore, the authors started to test several alloys for the conversion of CO_2. Indeed, it was demonstrated that methanol can be selectively obtained by the use of a CuNi alloy, whereas acetic acid and CO can be produced by alloying Cu with Sn or Pb.

Based on these results, several research groups improved the CO_2 reduction by using alloys with various ratios of different metals. Anagostopoulos and Kyriacou modified a copper surface with gold and studied the effect on the CE in contrast to gold content at the electrode [92]. By increasing the gold amount to 7.6% on the surface, the CE for the reduction of CO_2 to CO increased by a factor of 4, whereas the CE for the formation of hydrocarbon was significantly decreased. In particular, the formation of liquid products was depressed from 30% to 3% with higher amounts of gold (Figure 5.31).

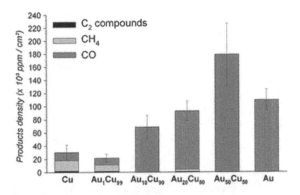

FIGURE 5.32 Ratios of gaseous products by electrolysis of CO_2 at copper, gold, and Cu–Au alloys. (With permission from Christophe, J. et al., *Electrocatalysis*, 3, 139, 2012.)

FIGURE 5.33 (a) Scanning tunneling microscopy (STM) of nano-porous copper, (b) STM of nano-porous Au–Cu alloy, and (c) STM of a bulk Cu–Au alloy.

In the recent years, this concept of copper–gold surfaces was adopted by Zhang et al. [93] and Buess-Herman et al. [94]. In contrast to the previous work, well-defined alloys and surfaces were synthesized and subjected to CO_2 reduction, since it was reported that a defined copper surface plays the major role. The authors focused on the formation of gases (CO, CH_4, C_2H_4, and C_2H_6) at single crystalline Cu, Au, Au_1Cu_{99}, $Au_{10}Cu_{90}$, $Au_{20}Cu_{80}$, or $Au_{50}Cu_{50}$ surfaces [94]. $Au_{50}Cu_{50}$ was identified to be the best alloy for the reduction of CO_2 to CO since the highest selectivity and the largest fraction of CO were detected and the Faradaic efficiency was increased from 12% up to 20% (Figure 5.32).

In contrast to Buess-Herman, Zhang et al. established nanostructured Cu–Au alloys for the reduction of CO_2 to formate, methanol, and ethanol in aqueous solution [93]. The authors prepared three alloys with different gold to copper ratios via electrodeposition of gold on nano-porous Cu films [95]. By the use of nano-porous materials, the surface is increased (Figure 5.33). This nano-porous material was much more efficient, which was identified by the comparison of the Faradic efficiencies of these alloys with bulk and nanostructured copper (Figure 5.34). The study revealed that $Cu_{63.9}Au_{36.1}$ exhibits a total faradic efficiency of 15.9% for methanol and 12% for ethanol, which is 19 times higher than for copper. Moreover, the electroanalytical studies proved that the alloy not only catalyzed CO_2 reduction but also CO reduction.

Beside Cu–Au alloys, Park et al. applied three lead-tin films on carbon paper for the CO_2 reduction and analyzed their potential for the conversion to formate in

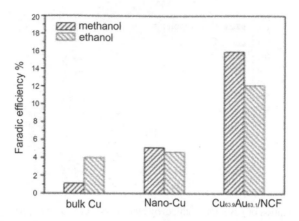

FIGURE 5.34 Faraday efficiency of copper vs. nano-porous copper and nano-porous Cu–Au alloy. NCF, nano-porous Cu film. (With permission from Jia, F. et al., *J. Power Sources*, 252, 85, 2014.)

competition to hydrogen evolution in aqueous media. All alloys were characterized prior to use via X-ray diffraction to determine their stoichiometry and by scanning electron microscopy (SEM) to assess morphology. Interestingly, the geometry and morphology of the alloy surface are quite different compared to that observed for the pure metals (Figure 5.35). The pure metals show crystals with tetragonal shape, whereas the alloy exists of crystals with a round habitus. To study their behavior in the reduction, electrolyses at a constant potential of $-2.0\,V$ vs. Ag/AgCl in the presence of CO_2 were run in a H-type divided cell (Figure 5.36). By a combination of both metals $(Sn_{56.3}Pb_{43.7})$, hydrogen evolution can be reduced significantly, whereas the production of $HCOO^-$ was increased simultaneously (Figure 5.36a). In addition, a higher current density could be successfully applied at the novel alloy system (Figure 5.36b).

Beside the reduction of CO_2, the application of novel cathode materials for electroorganic synthesis is only scarcely described. One impressive example is the application of the so-called $12CaO{\cdot}7Al_2O_3$ electride-type material. This electrode material is prepared by the substitution of free oxygen anions from the electric insulator $12CaO{\cdot}7Al_2O_3$ by electrons to achieve the conducting electride $[Ca_{24}Al_{28}O_{64}]^{4+}$ [97]. This electride exhibits two important characteristics which makes the material highly interesting as electrode in electrochemical conversions of a gaseous starting material; first, the high density of electrons from the electride, and second, a nano-porous structure (see Figure 5.37) [98]. Thereby, gaseous starting materials can be absorbed into the nano-porous structure and then be efficiently reduced.

The use of this cathode material was initially described by Li et al. [99]. The authors reduced O_2 to O_2^- which then transforms aryl boronic acids, like **77**, into the respective phenols **78** in excellent yield (Figure 5.38).

This concept was then applied onto the selective monocarboxylation of acrylates and styrene derivatives with CO_2 (Figure 5.39) [100]. The alkene can be reduced at the electrode to form the corresponding radical anion. This reactive species is then directly trapped by CO_2, which is absorbed at the porous electrode surface. Upon a second electron transfer a dianion is formed, which is then quenched by H^+ and

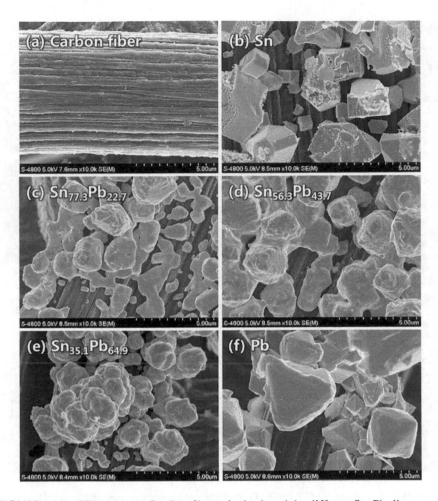

FIGURE 5.35 SEM images of carbon fibers, tin, lead, and the different Sn–Pb alloys.

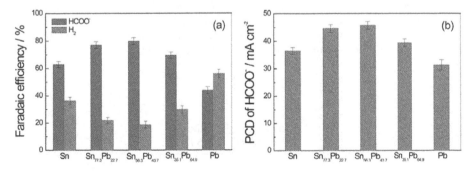

FIGURE 5.36 (a) Faradaic efficiencies for the CO_2 reduction to formate and hydrogen on tin, lead, and different tin–lead alloys. (b) Maximal current density during the electrolysis of CO_2 on different electrode materials. PCD, partial current density. (With permission from Choi, S.Y. et al., *ACS Sustain. Chem. Eng.*, 4, 1311, 2016.)

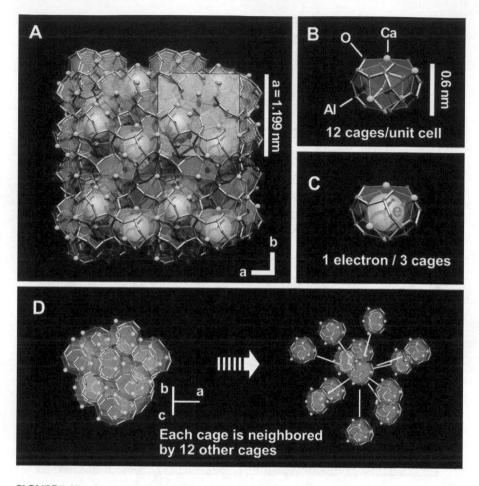

FIGURE 5.37   Structure of 12aO•7Al$_2$O$_3$ electride. (A) Crystal structure of [Ca$_{24}$Al$_{28}$O$_{64}$]$^{4+}$(4e$^-$), yellow region=a unit cell; (B) the framework of stoichiometric C12A7 is composed of 12 cages, 2 out of 12 cages are occupied by O^{2-} ions. (C) 4 out of 12 cages are occupied by electrons in place of O^{2-} ions. (D) Each cage is surrounded by 12 neighboring cages. When electrons are randomly distributed in the cages, 99% of the electrons have a neighboring electron. (With permission from Matsuishi, S. et al., *Science*, 301, 626, 2003.)

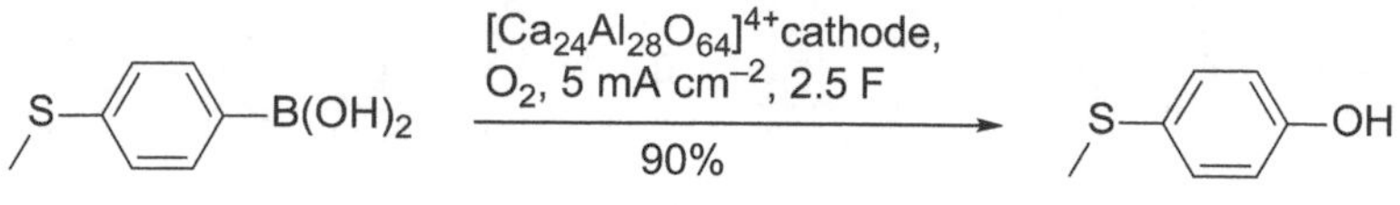

FIGURE 5.38   Superoxide-mediated synthesis of phenols from boronic acids at electride cathode. (With permission from Li, J. et al., *Electrochem. Commun.*, 17, 52, 2012.)

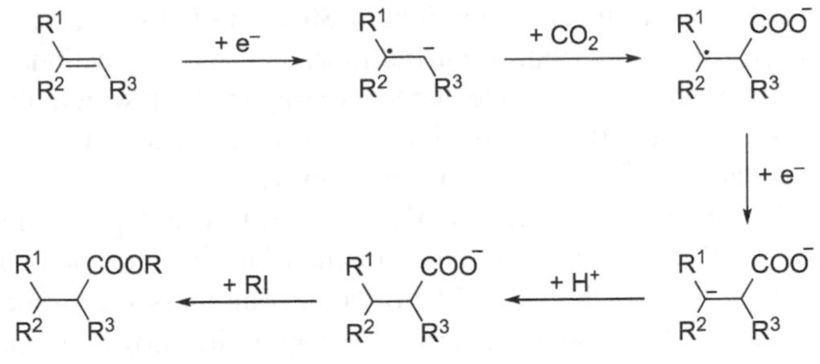

FIGURE 5.39 Carboxylation of alkenes at electride electrode (R^1=Ph, H; R^2=Me, H, Ph; R^3=Ph, H, Me, COOMe). (With permission from Li, J. et al., *Electrochem. Commun.*, 44, 45, 2014.)

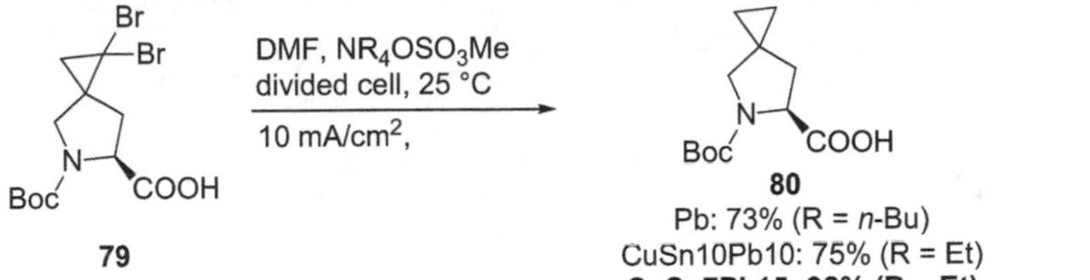

80
Pb: 73% (R = *n*-Bu)
CuSn10Pb10: 75% (R = Et)
CuSn7Pb15: 98% (R = Et)
Sn: 40% (R = Et)
Cu, Zn, BDD, graphite, GC: product mixture

FIGURE 5.40 Electrochemical dehalogenation of a pharmaceutically relevant proline derivative at a lead cathode.

alkyliodide within the work-up protocol. The major advantage of this novel electrode material is, that no sacrificial anodes are required [101–104]. Furthermore, only the mono carboxylated product is formed, whereas in other cases a product mixture of mono and double carboxylated products are obtained [105,106].

However, a significant disadvantage is the intricate preparation of the electride cathode and so far its limitation to gaseous starting materials. Another approach to establish stable and more powerful lead electrodes was found by screening for the twofold dehalogenation of 1,1-dibromo-cyclopropanes, especially for a pharmaceutically relevant proline derivative **79** (Figure 5.40). All common cathode material except lead exhibit an incomplete conversion and results in a mixture of difficult-to-separate products. Therefore, an alternative electrode material was required, which is easily available, inexpensive, and exhibits a higher chemical and mechanical stability as lead. It turned out that leaded bronze, an alloy of Cu, Sn, and Pb, fulfills all the requirements [22]. These alloys are commonly used in mechanical bearings and, therefore, were not brittle. Furthermore, they are easily available in metal trading shops in three different compositions ($CuSn_{10}Pb_{10}$, $CuSn_7Pb_{15}$, and $CuSn_5Pb_{20}$) at low costs (leaded bronze: 6–12 €/kg vs. lead: 2 €/kg).

The leaded bronzes with 10% and 15% lead were able to promote a full conversion of the starting material. In particular, $CuSn_7Pb_{15}$ is a highly potent material, since it provides almost quantitative yield and can be scaled up to 50 g of starting material [107]. However, the crude product of **80** showed almost no contamination with lead, copper, or tin (<10 ppm). Furthermore, late-stage functionalization of a cyclosporine A derivative **81** is viable, demonstrating the high selectivity of the conversion (Figure 5.41). Additionally, by fine tuning of the electrolyte also, a selective single dehalogenation to **82** is possible in excellent yield.

Actual studies on these materials reveal that these alloys are quite inhomogeneous with profound local deviations from their nominal bulk compositions. Broekmann et al. analyzed the samples by laser ablation/ionization mass spectrometry (LIMS) and prepared a three-dimensional composition map of the alloys (Figure 5.42). This

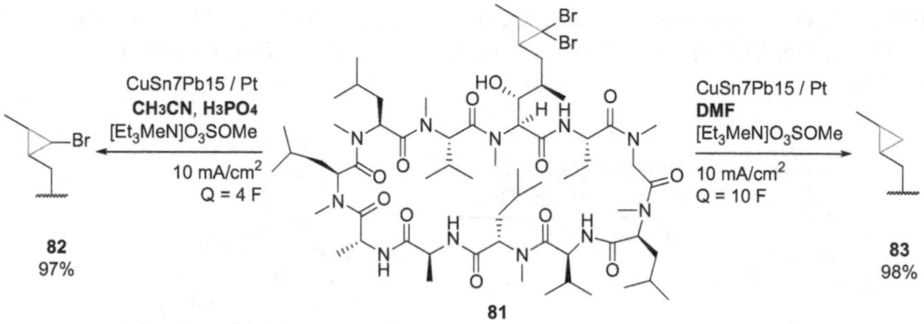

FIGURE 5.41 Selective late-stage functionalization of a cyclosporine A derivative via mono or twofold dehalogenation. (With permission from Gütz, C. et al., *Chem. Eur. J.,* 21, 13878, 2015.)

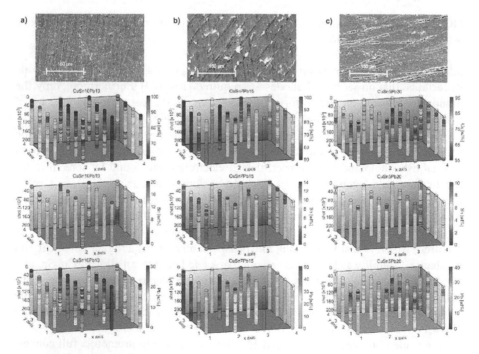

FIGURE 5.42 Results from LIMS measurement of the three different leaded bronze. (With permission from Grimaudo, V. et al., 2016, Unpublished results.)

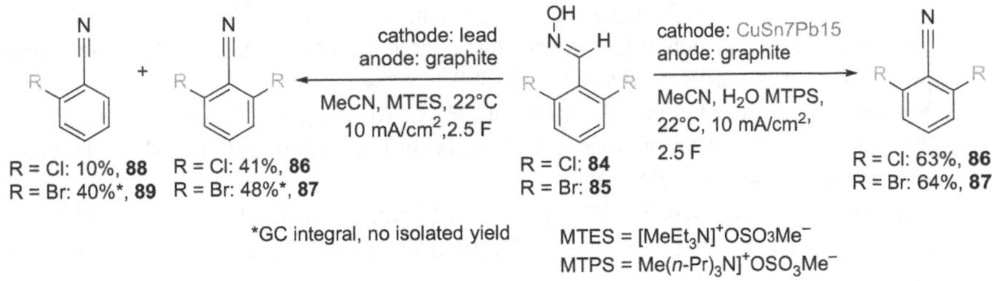
cathode: lead
anode: graphite
MeCN, MTES, 22°C
10 mA/cm2,2.5 F
cathode: CuSn7Pb15
anode: graphite
MeCN, H2O MTPS,
22°C, 10 mA/cm2,
2.5 F
R = Cl: 10%, 88
R = Br: 40%*, 89
R = Cl: 41%, 86
R = Br: 48%*, 87
R = Cl: 84
R = Br: 85
R = Cl: 63%, 86
R = Br: 64%, 87
*GC integral, no isolated yield
MTES = [MeEt3N]+OSO3Me−
MTPS = Me(n-Pr)3N]+OSO3Me−

FIGURE 5.43 Electrosynthesis of nitriles via a domino oxidation reduction sequence. (With permission from Hartmer, M.F. and Waldvogel, S.R., *Chem. Commun.*, 51, 16346, 2015; Gütz, C. et al., *ChemElectroChem.*, 5, 247, 2018.)

revealed local hot spots of lead in the alloy [108]. Additionally, in the CV measurements, a significantly broadened reduction peak is observed in contrast to the CV measurement of pure lead [22].

To investigate if this material is also suitable for other electroorganic conversions, we applied the leaded bronze in a recently published oxidation-reduction domino conversion of oximes into nitriles [109]. For the conversion at lead, we observed a dehalogenation of halogenated oximes **84** and **85** and a low yield of 41% for the dichloro derivative **86** and a difficult-to-separate product mixture for the dibromo derivative **87** (Figure 5.43).

The same conversion with leaded bronze as the electrode material and methyltripropyl ammonium methylsulfate as the supporting electrolyte provides the desired product **86** in 63% and **87** in 64% yield in the absence of the dehalogenated byproducts **88** and **89** [110].

5.5 SUMMARY

In summary, the first applications of leaded bronze in electrolyses indicate that this material offers a promising alternative for lead cathodes in electroorganic synthesis, which prevents contamination of products with lead. Finally, the stability, selectivity, and yield can be increased in comparison to lead as cathode material. In many applications, anodic as well as cathodic, BDD turned out as a superior electrode material. Since the organic synthesis is now going electric [111], and new concepts for value-added electrolysis products are elaborated [112], additional innovative electrode materials are needed.

REFERENCES

1. Ivandini, T. A., Einaga, Y., Honda, K. and Fujishima, A. 2005. *Diamond Electrochemistry.* Amsterdam: Elsevier.
2. Martinez-Huitle, C. A. and Brillas, E. 2011. *Synthetic Diamond Films. Preparation, Electrochemistry, Characterizationb, and Applications.* Hoboken, NJ: Wiley.
3. 2016; Available from: http://condias.de/de/home/startseite.html.
4. Waldvogel, S. R., Mentizi, S. and Kirste, A. 2012. Boron-doped diamond electrodes for electroorganic chemistry. *Top. Curr. Chem.* 320: 1–31.
5. Kirste, A., Schnakenburg, G., Stecker, F., Fischer, A. and Waldvogel, S. R. 2010. Anodic phenol–arene cross-coupling reaction on boron-doped diamond electrodes. *Angew. Chem. Int. Ed.* 49: 971–975.

6. Francke, R., Cericola, D., Kötz, R., Weingarth, D. and Waldvogel, S. R. 2012. Novel electrolytes for electrochemical double layer capacitors based on 1,1,1,3,3,3-hexafluoro propan-2-ol. *Electrochim. Acta* 62: 372–380.

7. Marselli, B., Garcia-Gomez, J., Michaud, P.-A., Rodrigo, M. A. and Comninellis, C. 2003. Electrogeneration of hydroxyl radicals on boron-doped diamond electrodes. *J. Electrochem. Soc.* 150: D79–D83.

8. Zollinger, D., Griesbach, U., Pütter, H. and Comninellis, C. 2004. Methoxylation of p-tert-butyltoluene on boron-doped diamond electrodes. *Electrochem. Commun.* 6: 600–604.

9. Sumi, T., Saitoh, T., Natsui, K., Yamamoto, T., Atobe, M., Einaga, Y. and Nishiyama, S. 2012. Anodic oxidation on a boron-doped diamond electrode mediated by methoxy radicals. *Angew. Chem. Int. Ed.* 51: 5443–5446.

10. Rodrigo, M. A., Michaud, P. A., Duo, I., Panizza, M., Cerisola, G. and Comninellis, C. 2001. Oxidation of 4-chlorophenol at boron-doped diamond electrode for wastewater treatment. *J. Electrochem. Soc.* 148: D60–D64.

11. Panizza, M., Michaud, P. A., Cerisola, G. and Comninellis, C. 2001. Anodic oxidation of 2-naphthol at boron-doped diamond electrodes. *J. Electroanal. Chem.* 507: 206–214.

12. Comminellis, C. 2010. *Electrochemistry for the Enviroment*. New York, NY: Springer-Verlag.

13. Martinez-Huitle, C. A. and Brillas, E. 2011. *The Wiley Series on Electrocatalysis and Electrochemistry*. Hoboken, NJ: Wiley.

14. Waldvogel, S. R. and Elsler, B. 2012. Electrochemical synthesis on boron-doped diamond. *Electrochim. Acta* 82: 434–443.

15. Malkowsky, I. M., Rommel, C. E., Wedeking, K., Fröhlich, R., Bergander, K., Nieger, M., Quaiser, C., Griesbach, U., Pütter, H. and Waldvogel, S. R. 2006. Facile and highly diastereoselective formation of a novel pentacyclic scaffold by direct anodic oxidation of 2,4-dimethylphenol. *Eur. J. Org. Chem.* 2006: 241–245.

16. Eberson, L., Persson, O. and Hartshorn, M. P. 1995. Detection and reactions of radical cations generated by photolysis of aromatic compounds with tetranitromethane in 1,1,1,3,3,3-hexafluoro-2-propanol at room temperature. *Angew. Chem. Int. Ed.* 34: 2268–2269.

17. Eberson, L., Hartshorn, M. P., Persson, O. and Radner, F. 1996. Making radical cations live longer. *Chem. Commun.* 2105–2112.

18. Eberson, L., Hartshorn, M. P. and Persson, O. 1995. 1,1,1,3,3,3-Hexafluoropropan-2-ol as a solvent for the generation of highly persistent radical cations. *J. Chem. Soc. Perkin Trans.* 2: 1735–1744.

19. Elsler, B., Wiebe, A., Schollmeyer, D., Dyballa, K. M., Franke, R. and Waldvogel, S. R. 2015. Source of selectivity in oxidative cross-coupling of aryls by solvent effect of 1,1,1,3,3,3-hexafluoropropan-2-ol. *Chem. Eur. J.* 21: 12321–12325.

20. Kashiwagi, T., Elsler, B., Waldvogel, S. R., Fuchigami, T. and Atobe, M. 2013. Reaction condition screening by using electrochemical microreactor: Application to anodic phenol-arene C,C cross-coupling reaction in high acceptor number media. *J. Electrochem. Soc.* 160: G3058–G3061.

21. Tafel, J. 1911. Ungesättigte bleialkyle. *Chem. Ber.* 44: 323–337.

22. Gütz, C., Selt, M., Bänziger, M., Bucher, C., Römelt, C., Hecken, N., Gallou, F., Galvão, T. R. and Waldvogel, S. R. 2015. A novel cathode material for cathodic dehalogenation of 1,1-dibromo cyclopropane derivatives. *Chem. Eur. J.* 21: 13878–13882.

23. Edinger, C., Grimaudo, V., Broekmann, P. and Waldvogel, S. R. 2014. Stabilizing lead cathodes with diammonium salt additives in the deoxygenation of aromatic amides. *ChemElectroChem* 1: 1018–1022.

24. Yanson, A. I., Antonov, P. V., Yanson, Y. I. and Koper, M. T. M. 2013. Controlling the size of platinum nanoparticles prepared by cathodic corrosion. *Electrochim. Acta* 110: 796–800.

25. Yang, N., Waldvogel, S. R. and Jiang, X. 2016. Electrochemistry of carbon dioxide on carbon electrodes. *ACS Appl. Mater. Interfaces* 8:28357–28371.

26. Reufer, C., Hateley, M., Lehmann, T., Weckbecker, C., Sanzenbacher, R. and Bilz, J. 2006. EP 1 631702.

27. Lehmann, E. D. T. 2009. Personal Communication.

28. Nakata, K., Ozaki, T., Terashima, C., Fujishima, A. and Einaga, Y. 2014. High-yield electrochemical production of formaldehyde from CO_2 and seawater. *Angew. Chem. Int. Ed.* 53: 871–874.

29. Kojima, T., Obata, R., Saito, T., Einaga, Y. and Nishiyama, S. 2015. Cathodic reductive coupling of methyl cinnamate on boron-doped diamond electrodes and synthesis of new neolignan-type products. *Beilstein J. Org. Chem.* 11: 200–203.

30. Seo, B.-I., Wall, L. K., Lee, H., Buttrum, J. W. and Lewis, D. E. 1993. An improved practical synthesis of isomerically pure 3-endo-(p-Methoxybenzyl)isoborneol. *Synth. Commun.* 23: 15–22.

31. Tafel, J. and Pfeffermann, E. 1902. Elektrolytische reduction von oximen und phenyl-hydrazonen in schwefelsaurer lösung. *Chem. Ber.* 35: 1510–1518.

32. Griesbach, U., Zollinger, D., Pütter, H. and Comninellis, C. 2005. Evaluation of boron doped diamond electrodes for organic electrosynthesis on a preparative scale. *J. Appl. Electrochem.* 35: 1265–1270.

33. Schopohl, M. C., Siering, C., Kataeva, O. and Waldvogel, S. R. 2003. Reversible enantiofacial differentiation of a single heterocyclic substrate by supramolecular receptors. *Angew. Chem. Int. Ed.* 42: 2620–2623.

34. Siering, C., Grimme, S. and Waldvogel, S. R. 2005. Direct assignment of enantiofacial discrimination on single heterocyclic substrates by self-induced CD. *Chem. Eur. J.* 11: 1877–1888.

35. Schopohl, M. C., Faust, A., Mirk, D., Fröhlich, R., Kataeva, O. and Waldvogel, S. R. 2005. Synthesis of rigid receptors based on triphenylene ketals. *Eur. J. Org. Chem.* 2005: 2987–2999.

36. Bomkamp, M., Siering, C., Landrock, K., Stephan, H., Fröhlich, R. and Waldvogel, S. R. 2007. Extraction of radio-labelled Xanthine derivatives by artificial receptors: Deep insight into the association behaviour. *Chem. Eur. J.* 13: 3724–3732.

37. Schade, W., Bohling, C., Hohmann, K., Bauer, C., Orghici, O., Waldvogel, S. R. and Scheel, D. 2007. Photonic sensors for security applications. *Photonik Int.* 1: 32–34.

38. Börner, S., Orghici, R., Waldvogel, S. R., Willer, U. and Schade, W. 2009. Evanescent field sensors and the implementation of waveguiding nanostructures. *Appl. Opt.* 48: B183–B189.

39. Schwartz, U., Großer, R., Piejko, K.-E. and Arlt, D. 1987. Optisch aktive (Meth)-acrylamide, polymere daraus, verfahren zu ihrer herstellung und ihre verwendung zur racematspaltung. DE 3532356 A1.

40. Bömer, B., Großer, R., Lange, W., Zweering, U., Köhler, B., Sirges, W. and Grose-Bley, M. 1997. Chirale stationäre Phasen für die chromatographische Trennung von optischen Isomeren. DE 19546136 A1.

41. Looft, J., Vössing, T., Ley, J., Backes, M. and Blings, M. 2008. Substituierte cyclopropancarbonsäure(3-methyl-cyclohexyl)amide als geschmacksstoffe. EP 1989944 A1.

42. Kulisch, J., Nieger, M., Stecker, F., Fischer, A. and Waldvogel, S. R. 2011. Efficient and stereodivergent electrochemical synthesis of optically pure menthylamines. *Angew. Chem. Int. Ed.* 50: 5564–5567.

43. Rodrigo, E. and Waldvogel, S. R. 2018. Very simple one-pot electrosynthesis of nitrones starting from nitro and aldehyde components. *Green Chem.* 20: 2013–2017.
44. Fardel, R., Griesbach, U., Pütter, H. and Comninellis, C. 2006. Electrosynthesis of trimethylorthoformate on BDD electrodes. *J. Appl. Electrochem.* 36: 249–253.
45. Zollinger, D., Griesbach, U., Pütter, H. and Comninellis, C. 2004. Electrochemical cleavage of 1,2-diphenylethanes at boron-doped diamond electrodes. *Electrochem. Commun.* 6: 605–608.
46. Weinberg, N. L. and Weinberg, H. R. 1968. Electrochemical oxidation of organic compounds. *Chem. Rev.* 68: 449–523.
47. Clauson-Kaas, N., Limborg, F. and Glens, K. 1952. Electrolytic methoxylation of furan. *Acta Chem. Scand.* 6: 531–534.
48. Reufer, C., Lehmann, R., Sanzenbacher, R. and Weckbecker, C. 2004. Verfahren zur anodischen Alkoxylierung von organischen substraten. WO 2004085710 A2.
49. Fuchigami, T. and Inagi, S. 2011. Selective electrochemical fluorination of organic molecules and macromolecules in ionic liquids. *Chem. Commun.* 47: 10211–10223.
50. Hou, Y., Higashiya, S. and Fuchigami, T. 1997. Electrolytic partial fluorination of organic compounds. 22.1 highly regioselective anodic monofluorination of oxindole and 3-oxo-1,2,3,4-tetrahydroisoquinoline derivatives: Effects of supporting fluoride salts and anode materials. *J. Org. Chem.* 62: 8773–8776.
51. Fuchigami, T. 1997. Personal communication.
52. Okino, F., Shibata, H., Kawasaki, S., Touhara, H., Momota, K., Nishitani-Gamo, M., Sakaguchi, I. and Ando, T. 1999. Electrochemical fluorination of 1,4-difluorobenzene using boron-doped diamond thin-film electrodes. *Electrochem. Solid-State Lett.* 2: 382–384.
53. Yoshida, K. 1979. Regiocontrolled anodic cyanation of nitrogen heterocycles. Pyrroles and indoles. *J. Am. Chem. Soc.* 101: 2116–2121.
54. Fuchigami, T., Nojima, H., Shimoda, T. and Atobe, M. 2003. Electroorganic synthesis using boron-doped diamond electrodes. *203rd ECS Meeting Abstract* 2604: Paris.
55. Bäumer, U.-S. and Schäfer, H. J. 2005. Cleavage of alkenes by anodic oxidation. *J. Appl. Electrochem.* 35: 1283–1292.
56. Hili, R. and Yudin, A. K. 2006. Making carbon-nitrogen bonds in biological and chemical synthesis. *Nat. Chem. Biol.* 2: 284–287.
57. Bowie, A. L., Hughes, C. C. and Trauner, D. 2005. Concise synthesis of (±)-rhazinilam through direct coupling. *Org. Lett.* 7: 5207–5209.
58. Sousa, M. M., Melo, M. J., Parola, A. J., Morris, P. J. T., Rzepa, H. S. and de Melo, J. S. S. 2008. A study in mauve: Unveiling Perkin's dye in historic samples. *Chem. Eur. J.* 14: 8507–8513.
59. Yella, A., Lee, H.-W., Tsao, H. N., Yi, C., Chandiran, A. K., Nazeeruddin, M. K., Diau, E. W.-G., Yeh, C.-Y., Zakeeruddin, S. M. and Grätzel, M. 2011. Porphyrin-sensitized solar cells with cobalt (II/III)–based redox electrolyte exceed 12 percent efficiency. *Science* 334: 629–634.
60. Bikker, J. A., Brooijmans, N., Wissner, A. and Mansour, T. S. 2009. Kinase domain mutations in cancer: Implications for small molecule drug design strategies. *J. Med. Chem.* 52: 1493–1509.
61. Morofuji, T., Shimizu, A. and Yoshida, J.-i. 2013. Electrochemical C–H amination: Synthesis of aromatic primary amines via N-arylpyridinium ions. *J. Am. Chem. Soc.* 135: 5000–5003.
62. Waldvogel, S. R. and Möhle, S. 2015. Versatile electrochemical C–H amination via zincke intermediates. *Angew. Chem. Int. Ed.* 54: 6398–6399.

63. Herold, S., Möhle, S., Zirbes, M., Richter, F., Nefzger, H. and Waldvogel, S. R. 2016. Electrochemical amination of less-activated alkylated arenes using boron-doped diamond anodes. *Eur. J. Org. Chem.* 2016: 1274–1278.
64. von Nussbaum, F., Brands, M., Hinzen, B., Weigand, S. and Häbich, D. 2006. Antibacterial natural products in medicinal chemistry—Exodus or revival? *Angew. Chem. Int. Ed.* 45: 5072–5129.
65. Franke, R., Selent, D. and Börner, A. 2012. Applied hydroformylation. *Chem. Rev.* 112: 5675–5732.
66. Waldvogel Siegfried, R. 2010. Novel anodic concepts for the selective phenol coupling reaction. *Pure Appl. Chem.* p. 1055.
67. Barjau, J., Königs, P., Kataeva, O. and Waldvogel, S. R. 2008. Reinvestigation of highly diastereoselective pentacyclic spirolactone formation by direct anodic oxidation of 2,4-dimethylphenol. *Synlett* 2008: 2309–2312.
68. Barjau, J., Schnakenburg, G. and Waldvogel, S. R. 2011. Diversity-oriented synthesis of polycyclic scaffolds by modification of an anodic product derived from 2,4-dimethylphenol. *Angew. Chem. Int. Ed.* 50: 1415–1419.
69. Kirste, A., Nieger, M., Malkowsky, I. M., Stecker, F., Fischer, A. and Waldvogel, S. R. 2009. Ortho-selective phenol-coupling reaction by anodic treatment on boron-doped diamond electrode using fluorinated alcohols. *Chem. Eur. J.* 15: 2273–2277.
70. Kirste, A., Schnakenburg, G. and Waldvogel, S. R. 2011. Anodic coupling of guaiacol derivatives on boron-doped diamond electrodes. *Org. Lett.* 13: 3126–3129.
71. Kirste, A., Elsler, B., Schnakenburg, G. and Waldvogel, S. R. 2012. Efficient anodic and direct phenol-Arene C,C cross-coupling: The benign role of water or methanol. *J. Am. Chem. Soc.* 134: 3571–3576.
72. Elsler, B., Schollmeyer, D., Dyballa, K. M., Franke, R. and Waldvogel, S. R. 2014. Metal- and reagent-free highly selective anodic cross-coupling reaction of phenols. *Angew. Chem. Int. Ed.* 53: 5210–5213.
73. Wiebe, A., Schollmeyer, D., Dyballa, K. M., Franke, R. and Waldvogel, S. R. 2016. Selective synthesis of partially protected nonsymmetric biphenols by reagent- and metal-free anodic cross-coupling reaction. *Angew. Chem. Int. Ed.* 55: 11801–11805.
74. Lips, S., Wiebe, A., Elsler, B., Schollmeyer, D., Dyballa, K. M., Franke, R. and Waldvogel, S. R. 2016. Synthesis of meta-terphenyl-2,2″-diols by anodic C–C cross-coupling reactions. *Angew. Chem. Int. Ed.* 55: 10872–10876.
75. Little, R. D. and Moeller, K. D. 2002. Organic electrochemistry as a tool for synthesis: Umpolung reactions, reactive intermediates, and the design of new synthetic methods. *Electrochem. Soc.* 11: 36–42.
76. Mihelcic, J. and Moeller, K. D. 2003. Anodic cyclization reactions: The total synthesis of alliacol A. *J. Am. Chem. Soc.* 125: 36–37.
77. Sun, H. and Moeller, K. D. 2003. Building functionalized peptidomimetics: New electroauxiliaries and the use of a chemical oxidant for introducing N-acyliminium ions into peptides. *Org. Lett.* 5: 3189–3192.
78. Raess, P. W., Mubarak, M. S., Ischay, M. A., Foley, M. P., Jennermann, T. B., Raghavachari, K. and Peters, D. G. 2007. Catalytic reduction of 1-iodooctane by nickel(I) salen electrogenerated at carbon cathodes in dimethylformamide: Effects of added proton donors and a mechanism involving both metal- and ligand-centered one-electron reduction of nickel(II) salen. *J. Electroanal. Chem.* 603: 124–134.
79. Wolf, N. L., Peters, D. G. and Mubarak, M. S. 2006. Electrochemical reduction of 1-halooctanes at platinized platinum electrodes in dimethylformamide containing tetramethylammonium tetrafluoroborate. *J. Electrochem. Soc.* 153: E1–E4.

80. Peters, D. G. 2001. Halogenated organic compounds. In *Organic Electrochemistry*, ed. H. Lund and O. Hammerich, 341–377. New York: Marcel Dekker.

81. Beck, F. 1974. Kathodische spaltung der kohlenstoff-halogen-bindung. In *Elektroorganische Chemie*, ed. F. Beck, 183–186. Berlin: Akademie-Verlag.

82. Torii, S. 2006. Electroreduction of halogenated compounds. In *Electroorganic Reduction Synthesis*, ed. S. Torii, 331–337. Weinheim: Wiley-VCH Verlag GmbH & Co. KGaA.

83. Armstrong, R. D. and Bladen, K. L. 1977. The anodic dissolution of lead in oxygenated and deoxygenated sulphuric acid solutions. *J. Appl. Electrochem.* 7: 345–353.

84. Roll, K. H. 1953. *Ind. Eng. Chem.* 45: 2210–2214.

85. Stohs, S. J. and Bagchi, D. 1995. Oxidative mechanisms in the toxicity of metal ions. *Free Radic. Biol. Med.* 18: 321–336.

86. Samarin, K. M., Makarochkina, S. M., Tomilov, A. P. and Zhitareva, L. V. 1980. Cathodic synthesis of tetraethyllead. *Elektrokhimiya* 16: 326–331.

87. Fleischmann, M., Pletcher, D. and Vance, C. J. 1971. Reduction of alkyl halides at a lead cathode in dimethylformamide. *J. Electroanal. Chem. Interfacial Electrochem.* 29: 325–334.

88. Edinger, C. and Waldvogel, S. R. 2014. Electrochemical deoxygenation of aromatic amides and sulfoxides. *Eur. J. Org. Chem.* 2014: 5144–5148.

89. Edinger, C., Kulisch, J. and Waldvogel, S. R. 2015. Stereoselective cathodic synthesis of 8-substituted (1R,3R,4S)-menthylamines. *Beilstein J. Org. Chem.* 11: 294–301.

90. Rogelj, J., den Elzen, M., Höhne, N., Fransen, T., Fekete, H., Winkler, H., Schaeffer, R., Sha, F., Riahi, K. and Meinshausen, M. 2016. Paris agreement climate proposals need a boost to keep warming well below 2°C. *Nature* 534: 631–639.

91. Watanabe, M., Shibata, M., Katoh, A., Sakata, T. and Azuma, M. 1991. Design of alloy electrocatalysts for CO_2 reduction. *J. Electroanal. Chem. Interfacial Electrochem.* 305: 319–328.

92. Kyriacou, G. and Anagnostopoulos, A. 1992. An international journal devoted to all aspects of electrode kinetics, interfacial structure, properties of electrolytes, colloid and biological electrochemistry. Electrochemical reduction of CO_2 at Cu+Au electrodes. *J. Electroanal. Chem.* 328: 233–243.

93. Jia, F., Yu, X. and Zhang, L. 2014. Enhanced selectivity for the electrochemical reduction of CO_2 to alcohols in aqueous solution with nanostructured Cu–Au alloy as catalyst. *J. Power Sources* 252: 85–89.

94. Christophe, J., Doneux, T. and Buess-Herman, C. 2012. Electroreduction of carbon dioxide on copper-based electrodes: Activity of copper single crystals and copper–gold alloys. *Electrocatalysis* 3: 139–146.

95. Jia, F., Zhao, J. and Yu, X. 2013. Nanoporous Cu film/Cu plate with superior catalytic performance toward electro-oxidation of hydrazine. *J. Power Sources* 222: 135–139.

96. Choi, S. Y., Jeong, S. K., Kim, H. J., Baek, I.-H. and Park, K. T. 2016. Electrochemical reduction of carbon dioxide to formate on tin–lead alloys. *ACS Sustain. Chem. Eng.* 4: 1311–1318.

97. Kim, S.-W., Matsuishi, S., Miyakawa, M., Hayashi, K., Hirano, M. and Hosono, H. 2007. Fabrication of room temperature-stable 12CaO 7Al2O3 electride: A review. *J. Mater. Sci.—Mater. Electron.* 18: 5–14.

98. Matsuishi, S., Toda, Y., Miyakawa, M., Hayashi, K., Kamiya, T., Hirano, M., Tanaka, I. and Hosono, H. 2003. High-density electron anions in a nanoporous single crystal: [Ca24Al28O64]4+(4e-). *Science* 301: 626–629.

99. Li, J., Yin, B., Fuchigami, T., Inagi, S., Hosono, H. and Ito, S. 2012. Application of 12CaO•7Al$_2$O$_3$ electride as a new electrode for superoxide ion generation and hydroxylation of an arylboronic acid. *Electrochem. Commun.* 17: 52–55.

100. Li, J., Inagi, S., Fuchigami, T., Hosono, H. and Ito, S. 2014. Selective monocarboxylation of olefins at 12CaO·7Al$_2$O$_3$ electride cathode. *Electrochem. Commun.* 44: 45–48.
101. Silvestri, G., Gambino, S. and Filardo, G. 1986. Electrochemical carboxylation of aldehydes and ketones with sacrificial aluminum anodes. *Tetrahedron Lett.* 27: 3429–3430.
102. Tokuda, M., Kabuki, T., Katoh, Y. and Suginome, H. 1995. Regioselective synthesis of β,γ-unsaturated acids by the electrochemical carboxylation of allylic bromides using a reactive-metal anode. *Tetrahedron Lett.* 36: 3345–3348.
103. Senboku, H., Yoneda, K. and Hara, S. 2013. Regioselective electrochemical carboxylation of polyfluoroarenes. *Electrochemistry* 81: 380–382.
104. Zhao, S.-F., Wang, H., Lan, Y.-C., Liu, X., Lu, J.-X. and Zhang, J. 2012. Influences of the operative parameters and the nature of the substrate on the electrocarboxylation of benzophenones. *J. Electroanal. Chem.* 664: 105–110.
105. Wang, H., Zhang, K., Liu, Y.-Z., Lin, M.-Y. and Lu, J.-X. 2008. Electrochemical carboxylation of cinnamate esters in MeCN. *Tetrahedron* 64: 314–318.
106. Orsini, M., Feroci, M., Sotgiu, G. and Inesi, A. 2005. Stereoselective electrochemical carboxylation: 2-phenylsuccinates from chiral cinnamic acid derivatives. *Org. Biomol. Chem.* 3: 1202–1208.
107. Gütz, C., Bänziger, M., Bucher, C., Galvão, T. R. and Waldvogel, S. R. 2015. Development and scale-up of the electrochemical dehalogenation for the synthesis of a key intermediate for NS5A inhibitors. *Org. Process Res. Dev.* 19: 1428–1433.
108. Grimaudo, V., Moreno-Gacia, P., Riedo, A., Meyer, S., Tulej, M., Neuland, M. B., Gütz, C., Waldvogel, S. R., Wurz, P. and Broekmann, P. 2016. Unpublished results.
109. Hartmer, M. F. and Waldvogel, S. R. 2015. Electroorganic synthesis of nitriles via a halogen-free domino oxidation-reduction sequence. *Chem. Commun.* 51: 16346–16348.
110. Gütz, C., Grimaudo, V., Holtkamp, M., Hartmer, M., Werra, J., Frensemeier, L., Kehl, A., Karst, U., Broekmann, P. and Waldvogel, S. R. 2018. Leaded bronze: An innovative lead substitute for cathodic electrosynthesis. *ChemElectroChem* 5: 247–252.
111. Wiebe, A., Gieshoff, T., Möhle, S., Rodrigo, E., Zirbes, M. and Waldvogel, S. R. 2018. Electrifying organic synthesis. *Angew. Chem. Int. Ed.* 57: 2–28; Wiebe, A., Gieshoff, T., Mohle, S., Rodrigo, E., Zirbes, M. and Waldvogel, S.R. Elektrifizierung der organischen synthese. *Angew. Chem.* 130, 2–30.
112. Möhle, S., Zirbes, M., Rodrigo, E., Gieshoff, T., Wiebe, A. and Waldvogel Siegfried, R. 2018. Modern electrochemical aspects for the synthesis of value-added organic products. *Angew. Chem. Int. Ed.* accepted [doi:0.1002/anie.201712732] Mohle, S., Zirbes, M., Rodrigo, E., Gieshoff, T., Wiebe, A. and Waldvogel, S. R. 2018. Moderne aspekte der elektrochemie zur synthese hochwertiger organischer produkte. *Angew. Chem.* accepted [doi:10.1002/ange.201712732].

6 Electrosynthesis of Functional Polymer Materials

Shinsuke Inagi and Naoki Shida
Tokyo Institute of Technology

CONTENTS

6.1 Introduction 127
6.2 Electrosynthesis of Conducting Polymers 128
    6.2.1 Electrochemistry of Conducting Polymers 128
    6.2.2 Anodic Electropolymerization of Aromatic Monomers 128
    6.2.3 Copolymer Synthesis by Electrochemical Methods 129
    6.2.4 Cathodic Electropolymerization of Aromatic Monomers 131
    6.2.5 Electrosynthesis of Polysilanes 131
6.3 Electrochemical Post-functionalization of Conducting Polymers 133
    6.3.1 Anodic Substitution Reaction of Conducting Polymers 133
    6.3.2 Cathodic Reaction and Paired Reactions 137
    6.3.3 Concurrent Reduction and Substitution Method 138
    6.3.4 Intramolecular Cyclization of Conducting Polymers 139
6.4 Electrosynthesis of Nonconjugated Polymers 140
    6.4.1 Electroactive Polymers 140
    6.4.2 Electrochemically Controlled Polymerization 141
    6.4.3 Electrochemical Cross-Linking Reaction of Nonconjugated
           Polymers 142
6.5 Conclusion and Outlook 142
References 143

6.1 INTRODUCTION

Organic electrosynthesis is a strong tool to produce a variety of functional compounds via electrochemically generated species as a key intermediate. It is possible to expand the methodology to polymer synthesis. Electrochemical polymerization utilizes electrogenerated species of a substrate for polymerization as a monomer or an initiator. In the former case, electrogenerated species of aromatic monomer couples with a polycondensation process to give a π-conjugated polymer, in which p-orbital of the carbon atoms on the aromatics overlaps throughout the polymer main chain. Such π-conjugated polymers are intrinsically conductive, thus called as conducting polymers [1,2]. Electrosynthetic methods are also available for activation of conducting polymers for further functionalization by using ionic species in the polymer chains. As for electrogenerated initiators for chain polymerization, ionic or radical initiators can be easily prepared at near surface of an electrode used and initiate chain polymerization of vinyl monomers. In both cases mentioned above, a variety of functionalities such as electroactivity, chromophore, and luminescence can be introduced into polymeric materials.

This chapter deals with the electrosynthesis of conducting polymers including electrooxidative polymerization, electroreductive polymerization, and electrosynthesis of polysilane. The concept of electrochemical functionalization of conducting polymers is described briefly. Since polymerization of vinyl monomers using electrogenerated initiator system has been summarized in another book [3,4], recent development in this process is enclosed here.

6.2 ELECTROSYNTHESIS OF CONDUCTING POLYMERS

6.2.1 ELECTROCHEMISTRY OF CONDUCTING POLYMERS

Conducting polymers with a narrow bandgap are redox-active, particularly in the thin-film state on a working electrode. The injection or removal of electrons to or from a conducting polymer results in the formation of polarons and bipolarons in the repeating structure (commonly known as doping), producing considerable variations in the physical properties of the polymer itself and imparting interesting features such as drastic color changes and electrical conductivity [5,6]. The charges thus generated along the polymer must be compensated by the addition of neighboring ions (or dopants), and the insertion and release of such dopants may induce volume changes in the conducting polymer (Scheme 6.1). Such a doping method can be categorized into either chemical or electrochemical in nature, and the latter is, of course, the most important from an electrochemical viewpoint. When an appropriate potential is applied to a conducting polymer film on an electrode, it can easily induce electron transfer at the electrode surface. In response, the doped state of the polymer causes changes in color, conductivity, and volume. In contrast, the application of the opposite potential to the doped polymer returns it to its neutral state. Such reversible switching of physical properties has applications to electrochromic devices, conducting materials, and actuators [7].

6.2.2 ANODIC ELECTROPOLYMERIZATION OF AROMATIC MONOMERS

Electron-rich aromatic and heteroaromatic monomers such as benzene, fluorene, pyrrole, and thiophene can be oxidized on an anode surface to form their radical cation states. The radical cations couple to form a carbon–carbon bond and, following deprotonation, lead to afford a neutral dimer (Scheme 6.2). The generated conjugated dimer has a lower oxidation potential than that of a monomer, so that further oxidation of the dimer leads to the oligomerization and polymerization. Although this mechanism is widely accepted, this mechanism is still controversial. The resulting highly conjugated polymers are no longer soluble in electrolytic medium to be

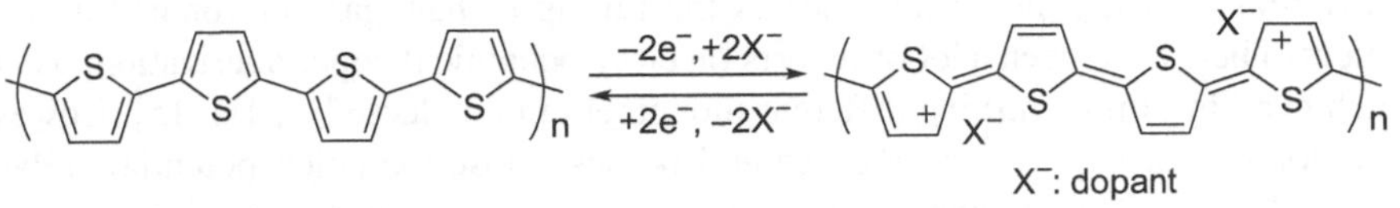

SCHEME 6.1 Electrochemical doping/dedoping of polythiophene.

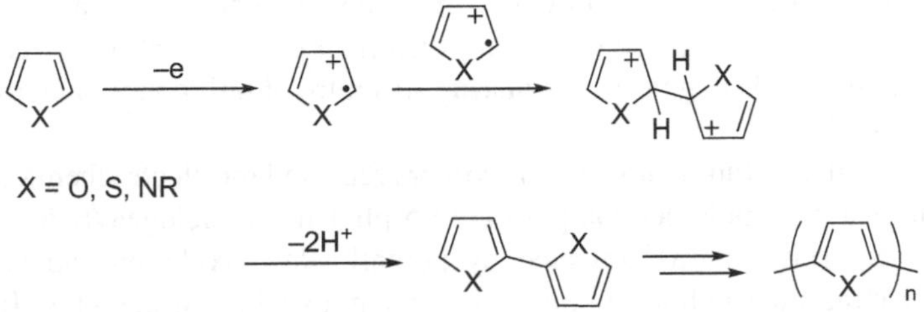

SCHEME 6.2 General scheme for electropolymerization of heteroaromatic monomers.

deposited on the anode surface as a film. The deposited conducting polymer is oxidatively doped during the application of potential for polymerization; thus, the film formed is not passive but still conductive to carry out continuous electrochemical reaction on the electrode. In this procedure, the coupling position of monomer can be predicted by spin density distribution on the aromatic ring [8]. An aromatic monomer having a functional group can yield the corresponding functional polymer by electropolymerization. The appropriate molecular design of monomers leads to the design of functionality of conjugated polymers.

In addition to conventional electrolytic solutions such as supporting salt/solvents, room temperature ionic liquids are known to be suitable for polymerization media [9]. There are several advantages in electropolymerization in ionic liquids: (i) electropolymerization proceeds rapidly, (ii) compatibility of a product film with an electrode surface is good, and (iii) product polymers show dense and smooth surfaces, high electric conductivity, high electrochemical capacitance, and good reversibility of redox cycles.

Recently, boron trifluoride-ether complexes are attractive as a new type of polymerization medium [10]. Boron trifluoride acts as a Lewis acid to form a complex with an aromatic monomer, which results in reducing its oxidation potential. Therefore, this method is effective for polymerization of aromatic monomers having relatively high oxidation potential. The reduction of polymerization potential can avoid overoxidation of product polymers.

6.2.3 COPOLYMER SYNTHESIS BY ELECTROCHEMICAL METHODS

Polymers consisting of two or more different monomer units, so-called copolymers, are of great importance because they possess various functionalities derived from each monomer component. For example, the combination of monomers with

different electrical properties enables the tuning of bandgaps of conducting polymers. There have been a lot of reports on electrochemical copolymerization, usually
achieved by simply mixing different monomers in an electrolyte [11–15]. However,
monomers for copolymerization should possess close oxidation potentials relative
to each other. The obtained polymer films show intermediate optical properties of
each homopolymer, which can be tuned by the feeding ratios of the monomers. To
produce alternating copolymers, one approach is to prepare a monomer composed of
two or more units beforehand, followed by electrochemical polymerization.

Dithienyl-substituted aromatics are candidates as monomers for alternating copolymers. As shown in Scheme 6.3, electrochemical polymerization of such dithienyl
derivatives proceeds to give an alternating structure of bithiophene and the aromatic
center [16–18].

In this context, Ruhlmann and co-workers reported the electrochemical synthesis
of an alternating copolymer composed of porphyrin and viologen (Scheme 6.4) [19].
First, zinc-porphyrin was activated by potentiostatic oxidation and reacted with
4,4'-bipyridine nucleophile, giving a porphyrin–pyridinium complex. In this step,
the obtained one-substituted adduct was not overoxidized due to the electron-
withdrawing nature of the bipyridinium group. This compound was regarded as
an AB-type monomer. This monomer was subjected into potential sweep electrolysis. The reduction wave derived from the generated viologen spacer increased

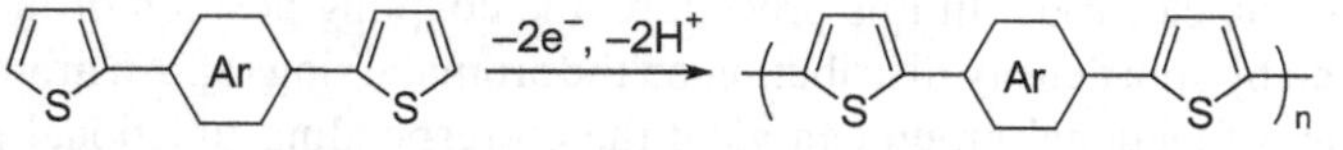

SCHEME 6.3 Electrochemical polymerization of dithienyl derivative.

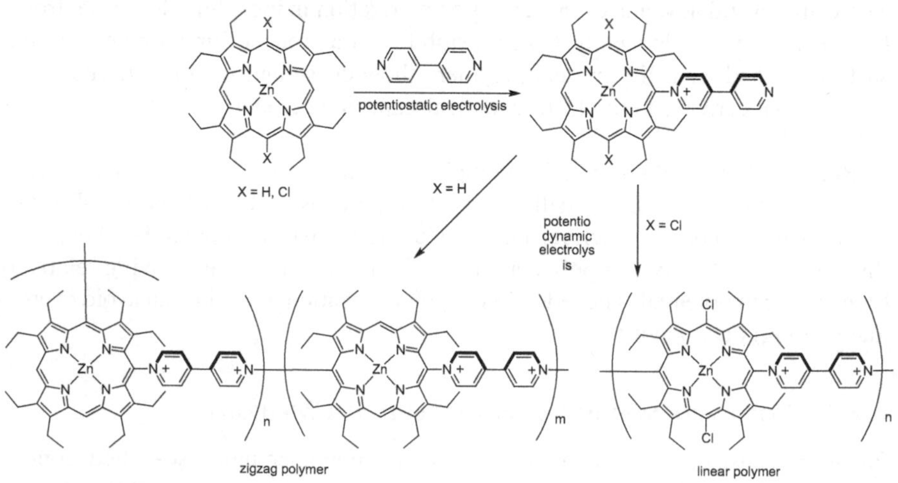

SCHEME 6.4 Electrochemical copolymerization of porphyrin and viologen.

continuously with the potential scanning cycles. This result strongly suggested the formation of a polymeric material possessing viologen unit on the working electrode via repetitive reactions of the AB-type monomer. Interestingly, when all *meso*-position of zinc porphyrin was not substituted (X = H), the polymer showed a more aggregated structure as determined by scanning electron microscopy and by atomic force microscopy, while the dichloro-substituted one (X = Cl) showed a linear fiber-like structure. The former one was derived from the statistical substitution of viologen moieties to porphyrin, which seemed to result in a zigzag polymer structure. The latter structure had a length of submicrometers and diameter of about 20 Å. This value corresponded to the molecular diameter of octaethylporphyrin (19 Å).

6.2.4 CATHODIC ELECTROPOLYMERIZATION OF AROMATIC MONOMERS

Aromatic dihalides are representative monomers for electroreductive polymerization. Cathodic reduction of an aromatic dihalide generates radical anion, followed by polymerization accompanying elimination of halide (Scheme 6.5). The bond formation takes place at the specific position of the monomer, where the halogen atom is substituted.

On electroreductive polymerization of 1,4-dibromopyridine, addition of catalytic amount of nickel(II) complex is effective and the complex works as mediator. Electrogenerated nickel(0) promotes dehalogenation of the dibromopyridine, followed by its polymerization [20]. Electrooxidative polymerization of pyridine is difficult because of its high oxidation potential; however, this reductive polymerization enables to produce polypyridines.

There are some reports on the synthesis of conducting polymer via electrochemical reduction. Utley and Gruber developed the reductive synthesis of poly(*p*-phenylene vinylene) (PPV) derivatives by the electrochemical reduction of a *p*-xylene derivative (PX). The reaction mechanism is summarized in Scheme 6.6 [21]. At first, two-electron reduction of PX gives quinodimethane, a reactive intermediate, and followed by polymerization to give poly(*p*-xylene) (PPX) when 2 F/mol of cathodic current was passed. This polymerization step is supposed to be radical mechanism evidenced by the meaningful effect of a radical-trapping reagent. When 4 F/mol was passed for PX, further debromination occurs to afford PPV. The reaction conditions were improved by utilizing an electrochemical mediator and succeeded to produce various PPV derivatives via electrochemical reduction.

6.2.5 ELECTROSYNTHESIS OF POLYSILANES

Polysilanes composed of linearly connected Si–Si bonds are in the class of conducting polymers and show unique optical properties [22]. The Kipping method, a well-known synthetic method of polysilanes from a dichlorosilane, is powerful,

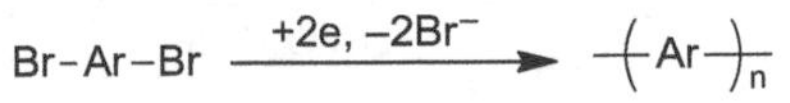

$$Br-Ar-Br \xrightarrow{+2e, -2Br^-} \left(Ar\right)_n$$

SCHEME 6.5   Electroreductive polymerization of aromatic dihalide.

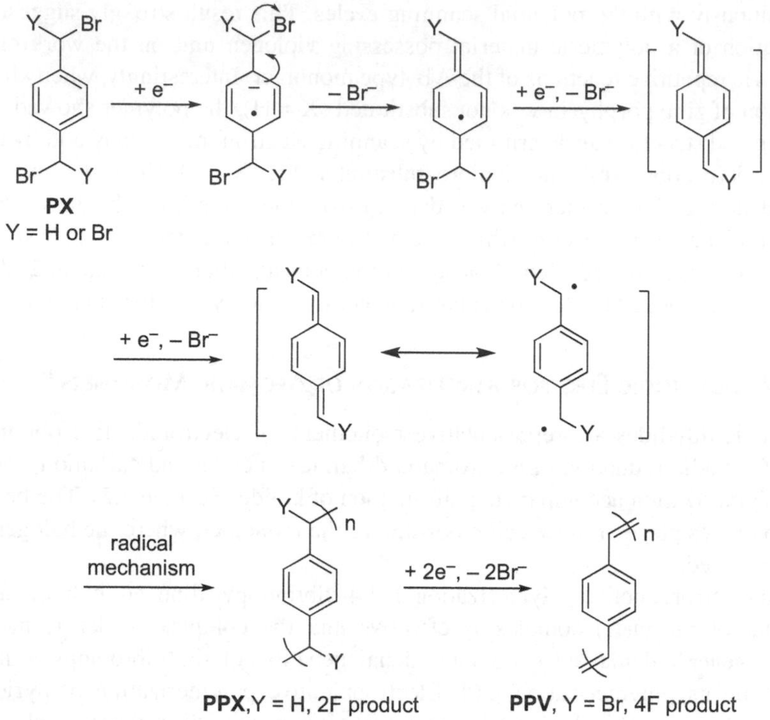

SCHEME 6.6   Electroreductive synthesis of PPV.

but the use of Na metal is indispensable. On the other hand, electrochemical reduction of chlorotrimethylsilane affords hexamethyldisilane without reducing reagent. Although polymerization of dichlorosilanes by electrochemical reduction is rather difficult, the use of Mg-sacrificial anode results in formation of polysilanes effectively (Scheme 6.7) [23]. The mechanism proposed is that Mg^{2+} once anodically dissolved from an anode is reduced at a cathode to give reactive Mg involving Si–Si bond formation of monomers. Cu and Al are also available as the sacrificial anode material for reductive synthesis of polysilanes.

Among the reductive polymerization of dichlorosilanes, a variety of copolymers such as random copolymers of silane monomers, poly(carbosilane)s, random copolymers of silane monomer, and germane monomer have been produced (Figure 6.1) [24,25].

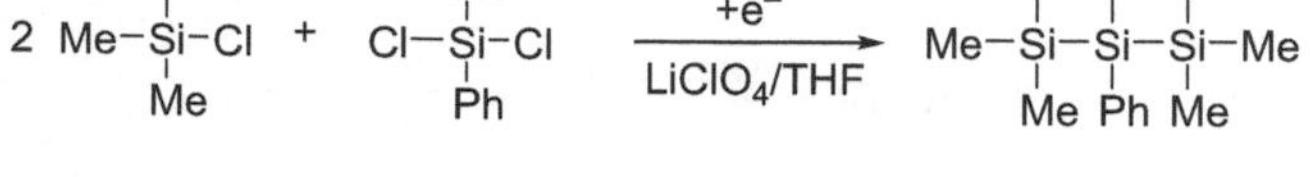

SCHEME 6.7   Electroreductive synthesis of polysilane.

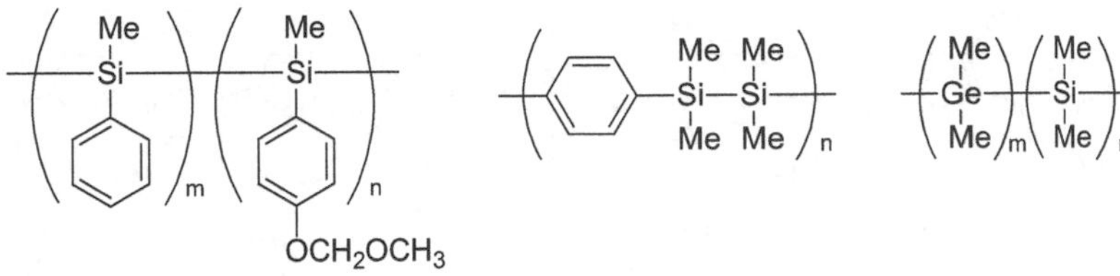

FIGURE 6.1 Copolymers based on polysilane.

6.3 ELECTROCHEMICAL POST-FUNCTIONALIZATION OF CONDUCTING POLYMERS

6.3.1 ANODIC SUBSTITUTION REACTION OF CONDUCTING POLYMERS

In the previous section, the contribution of electrochemistry to conducting polymer synthesis was described. Electrochemistry is very powerful to oxidize or reduce small molecules to generate reactive species or catalyst for polymerization. However, to address desired polymeric materials via electrochemistry, another approach is possible to use polymer reaction process, in which polymers are transformed into other structures. In this method, various polymers with different functionalities can be produced from a precursor polymer. Moreover, polymer reaction also enables the introduction of many functional groups, which may inhibit direct polymerization of monomers possessing such functional groups. From these points of view, this kind of polymer reaction (post-functionalization) is recognized as a strong methodology in polymer synthesis.

In the field of polymer chemistry, polymer reaction by electrochemical techniques has not been actively investigated because many polymers work as an insulator. However, conducting polymers can involve electron transfer at the electrode surface, and thus electrochemical post-functionalization is potentially realized. Conducting polymers can easily be oxidized to form doped state, possessing polarons or bipolarons in the main chain. The generated cation species are generally stabilized in the conjugated backbones with non-nucleophilic dopants but are reactive with nucleophiles to undergo nucleophilic substitution reactions on the conducting polymers. Such electrochemical polymer reactions enable fine-tuning of the degree of reaction manipulated by the charge passed in solid state (allowing simpler work-up processing). Therefore, electrochemical polymer reaction has attracted much attention as a novel way of post-functionalization of conducting polymers [26].

Pickup and co-workers reported pioneering works on electrochemical polymer reactions. They focused on the degradation mechanism of a conducting polymer, poly(3-methylthiophene) (P3MT), by nucleophilic reaction of water (Scheme 6.8) [27]. In this case, the resulting polymer undergoes isomerization and becomes its nonconducting form. On the other hand, the substitution reaction with chloride ions would give chlorinated P3MT having conductivity (Scheme 6.9). P3MT prepared on an electrode by electropolymerization was used as a substrate for the anodic substitution in the presence of tetraethylammonium chloride (Et$_4$NCl) as a supporting electrolyte and chlorine source [28,29]. The cyclic voltammogram (CV) of P3MT

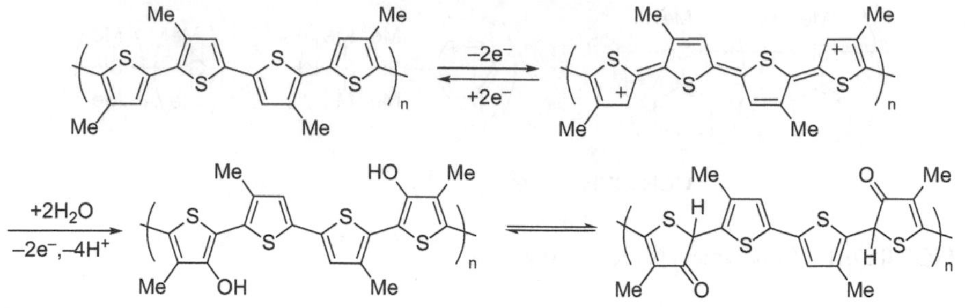

SCHEME 6.8 Degrading mechanism of poly(3-methylthiophene) in the presence of water.

showed a large oxidation current in comparison to CV measured with tetraethylam-monium perchlorate (Et$_4$NClO$_4$), a non-nucleophilic salt. Additionally, the oxidation process was irreversible. These results indicate that the following chemical reactions occurred as well as anodic doping. X-ray analysis evidenced the existence of C–Cl covalent bonds in the P3MT film, where the nucleophilic substitution reaction of Cl⁻ proceeded at the 4-position of the repeating thiophene rings (Scheme 6.9). The CV of the obtained chlorinated P3MT film in Et$_4$NClO$_4$/CH$_3$CN showed reversible redox waves derived from doping and dedoping similar to those observed in unmodified P3MT, while the oxidation potential of chlorinated P3MT slightly shifted anodi-cally. This was thought to be the electron-withdrawing effect of the chlorine substitu-ent. They also reported bromination and alkoxylation based on the same mechanism; however, iodination did not proceed due to its poor nucleophilicity. In addition, they reported the electrochemical polymer reaction using polypyrrole as a substrate. This post-functionalization is a powerful methodology in terms of not only realizing the facile tuning of electron density of conducting polymers but also the synthesis of methoxy-substituted polythiophenes, which cannot be obtained by electropolymer-ization of methoxythiophene monomer.

Bélanger and co-workers reported electrochemical chlorination of poly(3-(4-fluorophenyl)thiophene) in a similar manner to that of P3MT [30]. They also demonstrated the complete recovery of the original starting material following the application of a negative potential (–2 V vs. Ag/Ag⁺) due to one-electron reduction

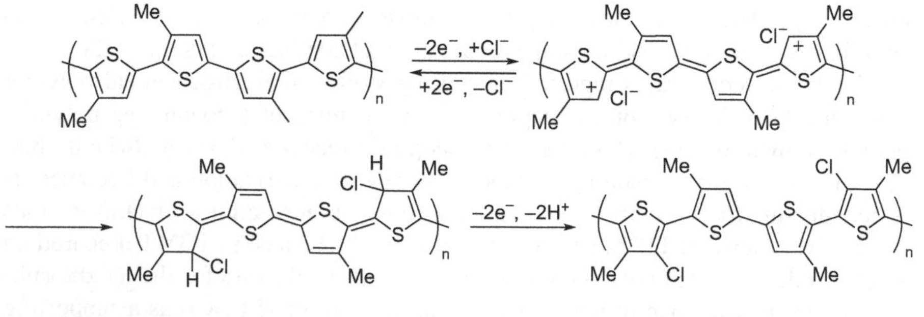

SCHEME 6.9 Electrochemical doping and following nucleophilic substitution by chloride ion.

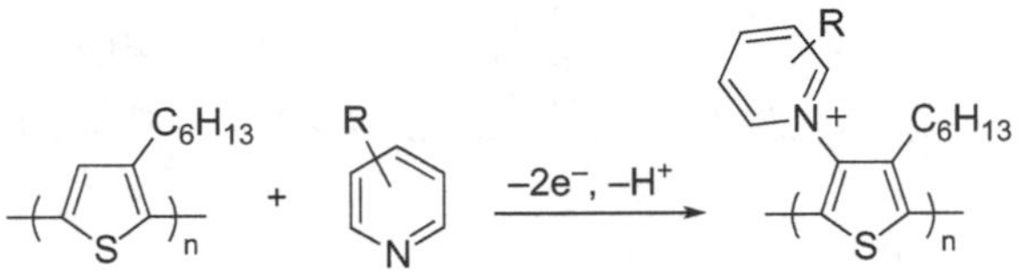

R = H, acetyl, methyl pyridinium

SCHEME 6.10 Electrochemical introduction of pyridinium moiety into poly (3-hexylthiophene).

and H-atom transfer from the electrolyte. Iyoda and co-workers reported electrochemical substitution of pyridine into the 4-position of poly(3-hexylthiophene) (Scheme 6.10) [31]. An electropolymerized poly(3-hexylthiophene) film on a Pt plate was subjected to anodic oxidation in the presence of pyridine derivatives (0.1 M). By constant potential or potential dynamic electrolysis, pyridinium moiety was successfully introduced, confirmed by Fourier-transform infrared spectroscopy and CV measurement. The introduction ratio of pyridinium moiety was achieved up to 60%.

Fabre and co-workers contributed to this field with anodic cyanation of poly(1,4-dimethoxybenzene) [32,33]. The anodic oxidation of electropolymerized poly(1,4-dimethoxybenzene) at 1.2 V vs. Ag/Ag$^+$ in the presence of tetraethylammonium cyanide (Et$_4$NCN) involved the aromatic substitution of hydrogen on the dimethoxybenzene unit rather than the replacement of aromatic methoxy group by CN (Scheme 6.11). The cyanation resulted in the linkage of ca. 1.4 CN group per dimethoxybenzene unit as evidenced by elemental analysis and X-ray photoelectron spectroscopy analysis.

In the researches mentioned above, characterization of post-functionalized polymers was conducted in solid state by infrared and ultraviolet–visible spectroscopy (UV–vis) measurements and energy-dispersive X-ray spectroscopy analysis. However, other analytic methods in solution such as nuclear magnetic resonance (NMR) spectroscopy, which is powerful for providing information on chemical structures, are not available. Inagi, Fuchigami, and co-workers recently developed the electrochemical polymer reactions with soluble conducting polymers [26].

An alternating copolymer composed of thiophene and 9,9-dioctylfluorene was prepared by Suzuki-Miyaura polycondensation reaction (Scheme 6.12). This poly(thiophene-*alt*-fluorene) (PTF) possesses a reactive site for oxidative halogenation (thiophene unit) and a solubilizing moiety (fluorene with long alkyl chains). Thus, the chemical structure of PTF was determined by NMR in chloroform-*d*, and

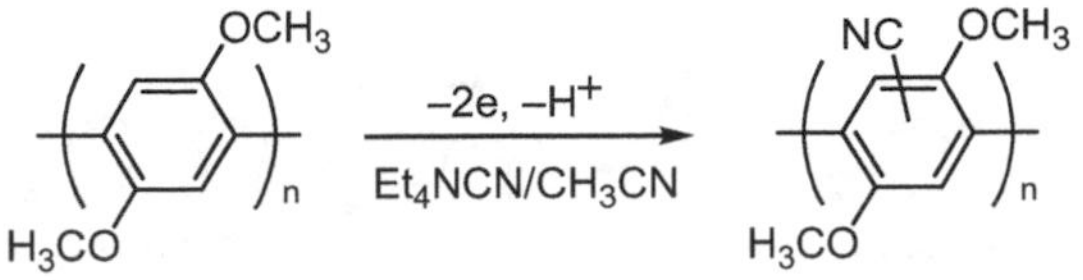

SCHEME 6.11 Electrochemical cyanation of poly(1,4-dimethyoxybenzene).

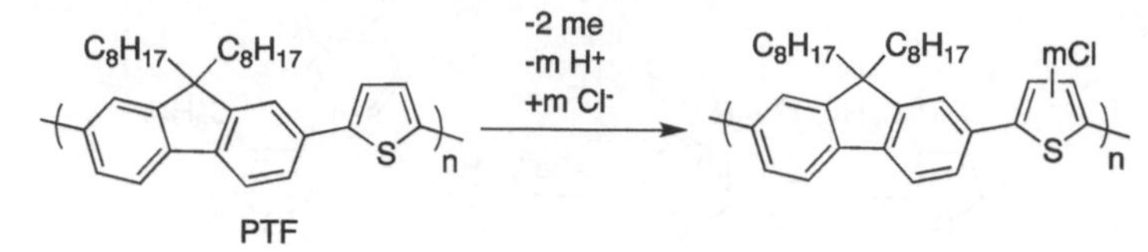

PTF

SCHEME 6.12   Electrochemical chlorination of a fluorene-thiophene-based copolymer.

the number-average molecular weight (M_n) and the polydispersity index (M_w/M_n) of PTF were estimated as 8700 and 2.1, respectively, by gel permeation chromatography. The cast-coated film from the solution of PTF on a Pt plate was used as an anode and oxidized by constant current technique in an electrolyte containing 0.1 M Et$_4$NCl in MeCN (poor solvent for PTF) [34]. After a passage of 30 F/mol of electricity, the film fixed on the anode was washed with methanol to remove supporting salts, and dissolved in chloroform-d for^1 H NMR analysis. The degree of chlorination was estimated by^1 H NMR. M_n and M_w/M_n of the product polymer were 9,000 and 2.3, respectively. These results strongly suggest that the electrochemical chlorination of the polymer film proceeded selectively at thiophene ring without side reactions like chain dissociation or cross-linking. It was also revealed that the amount of chlorine introduced could be controlled by changing the charge passed. By passing a current from 5, 10, and 20 F/mol for the PTF film, the degree of chlorination per thiophene unit, m, was determined as 0.66, 1.02, and 1.46, respectively [35]. The onset potentials of each film became more positive, indicating that the chlorinated PTF gained tolerance toward oxidation. As described here, the electrochemical polymer reaction is a good methodology to modify conducting polymers in a well-controlled manner, benefited by the facile manipulation of electrolytic conditions. Moreover, the halogenation of poly(fluorene-alt-bithiophene) [36] and poly(3-hexylthiophene) [37] was successfully accomplished.

The fluorination of organic compounds is one of the great challenges in organic synthesis toward application to pharmaceuticals, agrichemicals, and organic semiconductors. Anodic fluorination of organic molecules is one of the most successful processes, in which the electrochemical oxidation and subsequent nucleophilic fluorination are involved under mild conditions [38]. Based on this chemistry, the electrochemical fluorination of a conducting polymer film was carried out (Scheme 6.13). A copolymer composed of 9,9-dioctylfluorene and 9,9-diarylsulfanylfluorene (M_n = 6,400) was synthesized by coupling polymerization. Anodic fluorodesulfurization of the polymer proceeded almost completely by a passage of 24 F/mol of current in an electrolytic cell containing neat Et$_4$NF-5HF, evidenced by ^{1}H and ^{19}F NMR and elemental analysis [39,40]. Remarkably, the electrolysis of a nonconjugated

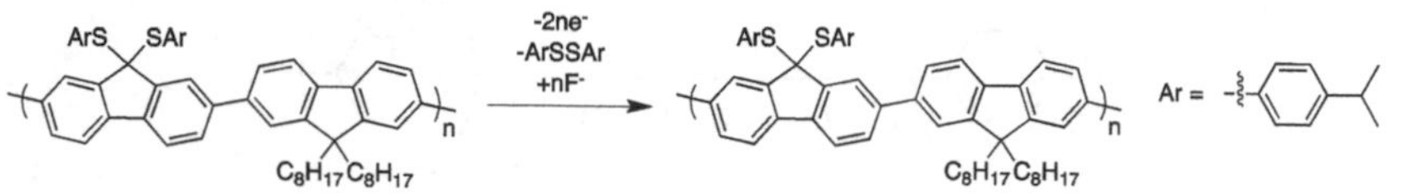

SCHEME 6.13   Electrochemical fluorination of a fluorene-based conducting polymer.

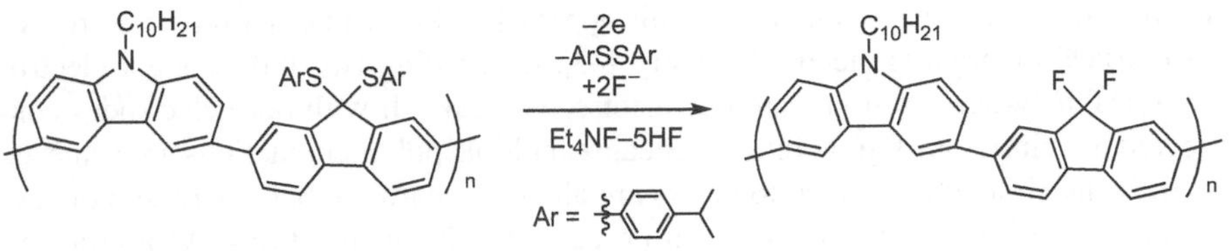

SCHEME 6.14 Anodic fluorodesulfurization of a polyfluorene derivative containing an alkylcarbazole moiety.

fluorene-based polymer did not afford the fluorinated product at all, indicating that the anodic doping of conjugated polyfluorene promoted the fluoride ion penetration into the polymer film as a dopant, and consequent substitution reaction proceeded.

A more powerful donor–acceptor type alternating copolymer consisting of 9-alkylcarbazole and 9,9-difluorofluorene was prepared by the electrochemical method similarly [41]. This was accomplished by electrolyzing a precursor polymer containing arylsulfanyl fluorene units in Et_4NF-5HF. Unfortunately, the conversion which was achieved under these conditions was only approximately 50% (Scheme 6.14); the oxidation of the alkylcarbazole units competed with that of the arylsulfanyl groups, which prevented effective fluorodesulfurization during electrolysis.

6.3.2 CATHODIC REACTION AND PAIRED REACTIONS

Such electrochemical polymer reactions can also be performed on a cathode initiated by cathodic doping. Poly(9,9-dioctylfluorene-*alt*-9-fluorenone) ($M_n = 4,100$) was prepared as a substrate, where the ketone group is possibly transformed into a methylene moiety by reductive hydrogenation. Electrolysis was carried out using tetraethylammonium p-toluenesulfonate (Et_4NOTs) as a supporting electrolyte and 2-propanol as a solvent and proton source (Scheme 6.15) [42]. A zinc cathode was effective for the reduction of ketone because of its high overpotential for reduction of proton. A passage of 16 F/mol of charge under a constant current condition resulted in the quantitative formation of the hydroxylated product polymer. The degree of reaction could be controlled by electricity passed and the lowest unoccupied molecular orbital level was also tunable depending on the progress of reaction.

The electrochemical polymer reactions described above have been carried out either on an anode or a cathode with a sacrificial reaction at a counter electrode. In order to improve current efficiency of the electrolytic system, the use of both electrodes

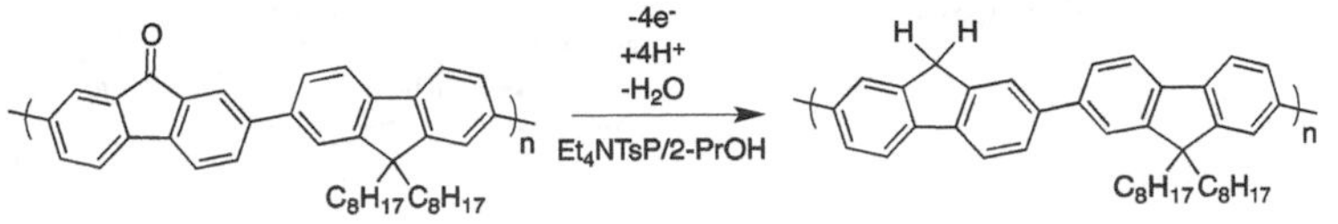

SCHEME 6.15 Electrochemical reduction of a ketone moiety in a fluorene-based conducting polymer.

for intended reactions is important. This approach is known as "paired" electrolysis in the field of organic electrochemistry [43]. Combining two independent electro-chemical polymer reactions into one electrochemical cell with both electrodes con-nected in a circuit is of great interest because multiple polymer reactions are achieved at the same time and product polymers are already separated. Such parallel polymer reactions are impossible in a solution process of polymer reactions. As a proof-of-concept of divergent-type paired reactions, the parallel electrochemical polymer reactions of single-precursor polymer, poly(9-fluorenol-*alt*-9,9-dioctylfluorene) **P1** were carried out in Et$_4$NOTs/2-propanol (Scheme 6.16). Two-electron oxidation of **P1** on a Pt anode afforded the **P2** having a 9-fluorenone moiety, whereas **P3**, a reduced form of **P1**, was obtained at a counter electrode (Zn) for the same electricity. The conversion of these reactions was almost quantitative by a passage of 16 F/mol of charge and the total current efficiencies of the system were doubled compared to the independent reactions. Interestingly, the emission color of the films of these polymers under UV irradiation was drastically different, namely blue for **P1**, no emission for **P2**, and yellow for **P3** [44].

6.3.3 CONCURRENT REDUCTION AND SUBSTITUTION METHOD

Polyaniline is one of the most known conducting polymers due to its multiple doping behaviors driven by oxidation and protonation. An emeraldine base (EB) of polyani-line is an oxidized state having the reactivity with nucleophiles. Han and co-workers developed the concurrent reduction and substitution (CRS) method, in which an EB of polyaniline can be reduced to be a leucoemeraldine base state along with the incorporation of various nucleophiles such as dialkylamines [45,46]. When an EB of polyaniline prepared by electropolymerization was immersed into a solution of tetrabutylammonium fluoride (Bu$_4$NF) in methanol, fluorine was introduced to the polyaniline main chain. Since only one fluorine atom could be introduced to one quinoidal structure, the repetitive treatments of the oxidation/CRS cycles provided the quantitative fluorination of the polyaniline (one fluorine atom per repeating ani-line) (Scheme 6.17) [47]. It was also found that the conductivity of the fluorinated polyaniline was higher than that of the precursor. Another approach to obtain the

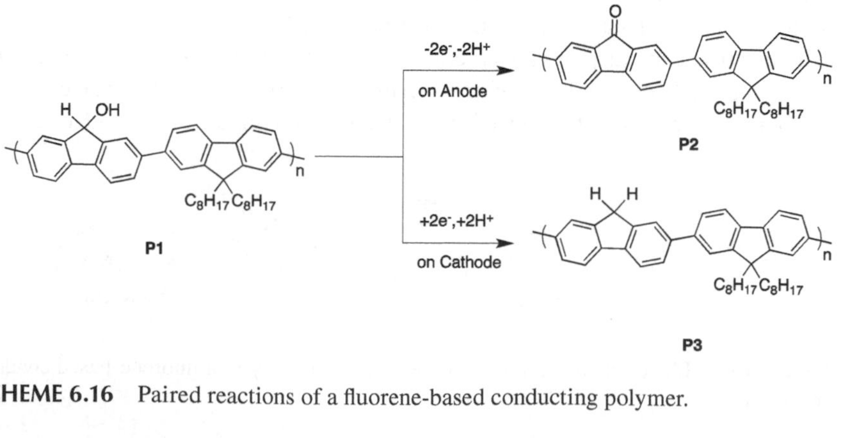

SCHEME 6.16 Paired reactions of a fluorene-based conducting polymer.

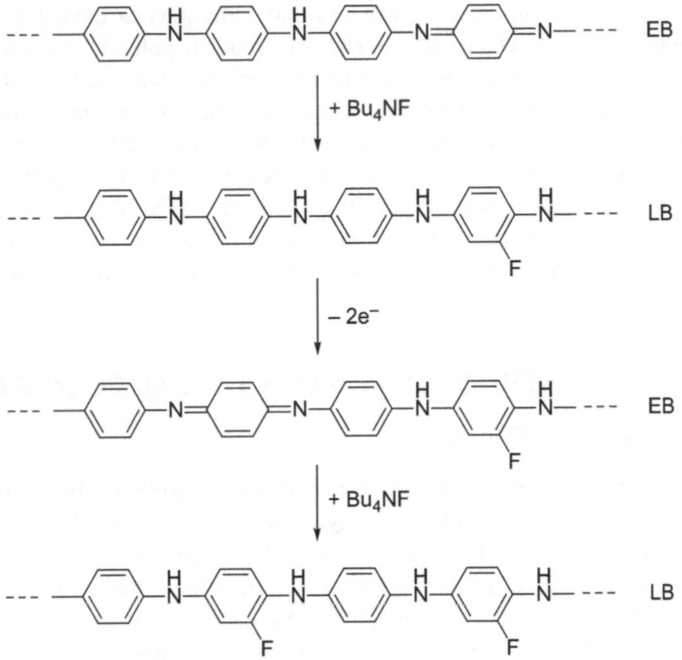

SCHEME 6.17   The CRS method for fluorination of polyaniline.

fluorinated polyaniline, i.e., electropolymerization of 2-fluoroaniline, did not give a high-molecular-weight polymer. Therefore, this method has a good advantage.

6.3.4   INTRAMOLECULAR CYCLIZATION OF CONDUCTING POLYMERS

Swager and co-workers reported the unique design of a segmented conducting polymer for oxidative cyclization in the polymer backbone (Scheme 6.18) [48]. They

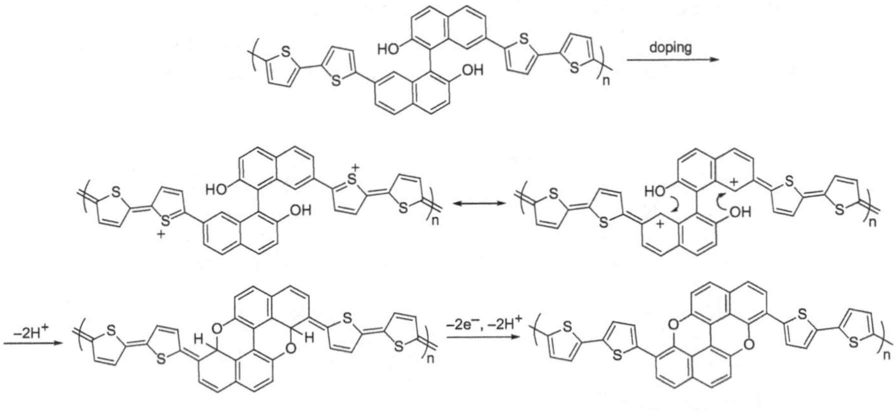

SCHEME 6.18   Electrochemical intramolecular cyclization of a binaphthol-containing conducting polymer.

prepared monomer composed of two terminal bithiophenes having a binaphthol center, followed by the electropolymerization under a potential sweep condition. Although the obtained segmented conducting polymer was stable and reversibly doped at a less positive potential range, it underwent irreversible cyclization at a highly anodic potential range to form a planar polycyclic heteroaromatic structure in the main chain. The in situ conductivity measurement of the polymers before and after the oxidative cyclization showed a clear difference. The conductivity onset shifted to a lower potential and maximum conductivity almost doubled. The UV–vis measurement also supported the formation of the desired product, exhibiting new optical transitions after the oxidation.

6.4   ELECTROSYNTHESIS OF NONCONJUGATED POLYMERS

6.4.1   ELECTROACTIVE POLYMERS

Conducting polymer films deposited on a working electrode in the course of electropolymerization do not inhibit following electrolysis because of their conductive nature. From this point of view, it is difficult to produce nonconjugated polymer films via electrode reactions because such a polymer film is assumed to passivate the working electrode. However, nonconjugated polymers bearing electroactive moieties exhibit conductivity arising from hopping of charges through the electroactive moieties in the polymer films.

Vinylpyridine complexes can be polymerized cathodically. Murray and co-workers reported a pioneering work in 1980s [49]. Zhong et al. have extended the ligand design and application of the polypyridine complexes [50]. Although the mechanism of this polymerization is not clear, it initiates by an anionic mechanism, and then propagates through a radical pathway. The monomers are composed of transition metals (Ru, Os, Fe, Co, Cr, and Ir) and ligands (pyridine, bipyridine, and terpyridine) with vinyl groups. As a typical example, a ruthenium complex $[Ru(dvbpy)(bpy)_2]^{2+}$ (bpy = 2,2'-bipyridine, dvbpy = 5,5'-divinyl-2,2'-bipyridine) was electropolymerized (Scheme 6.19) [51]. CV of this monomer showed an oxidation peak for Ru^{II} at +1.37 V, while three reduction peaks for the ligands were observed at −1.09, −1.40, and −1.63 V vs. Ag/AgCl. It was not until the potential scan reached the second reduction peak that the electropolymerization successfully proceeded. In the CV of

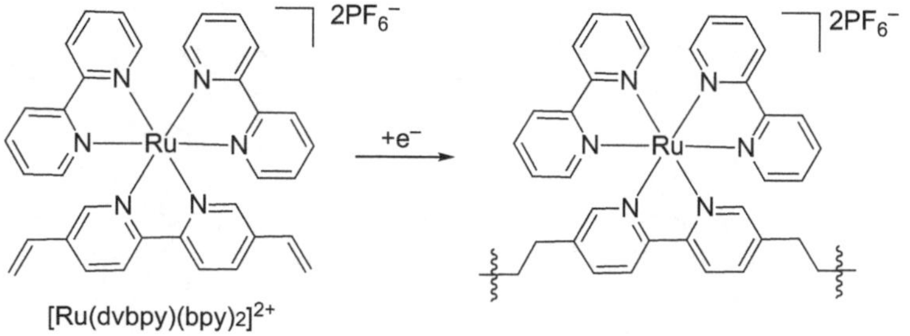

$[Ru(dvbpy)(bpy)_2]^{2+}$

SCHEME 6.19   Reductive electropolymerization of Ru-containing vinyl monomer.

the obtained film in a monomer-free electrolyte, the redox couple of $Ru^{II/III}$ appeared at almost the same potential compared to that of the monomer, indicating that the metal center was intact in the procedure. On the other hand, reduction potentials of the ligands negatively shifted after the polymerization due to the transformation of the vinyl moieties into saturated carbon–carbon bonds. Electroactive materials obtained by the polymerization procedure show interesting properties such as electrogenerated chemiluminescence, electrochromism, and electrocatalytic activity.

6.4.2 ELECTROCHEMICALLY CONTROLLED POLYMERIZATION

Electrogenerated reactive species of small molecules can work as polymerization initiators for ionic and radical polymerization of vinyl monomers [3,4]. For example, a methyl radical generated by Kolbe electrolysis of acetate can be an initiator for radical polymerization of vinyl monomers in electrolytic solution. Unfortunately, heterogeneity of initiator concentration in an electrochemical setup makes it difficult to obtain high-molecular-weight polymers.

Recent progress in living radical polymerization using transition metal catalyst (known as atom transfer radical polymerization, ATRP) is remarkable in both academia and industry [52,53]. These have been widely investigated for a couple of decades to provide a lot of complex polymeric materials like block copolymers, star polymers, polymer brushes, etc. In a living polymerization system, the chain length (degree of polymerization) could be controlled by changing the feed ratio of initiators and monomers.

As shown in Scheme 6.20, the equilibrium (k_a/k_{da}) between dormant species (A) and active species (B) determined by redox of Cu catalyst can tune the rate of monomer consumption, resulting in polymerization in a controlled manner. The external stimuli by electrochemical oxidation and reduction of the Cu catalyst give rise to change in the equilibrium and consequently more precise control of polymerization is possible (electrochemically mediated ATRP, eATRP) [54–56]. The eATRP process is excellent from not only an environmental aspect but also the new synthetic method with a facile control of polymerization. When a cathodic current is passed, Cu^I catalyst is generated to promote rapid polymerization, while in passing an anodic current, polymerization stops completely. This process is successfully repeated, and thus electrochemistry can manipulate the living polymerization.

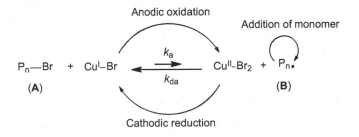

SCHEME 6.20 Mechanism of eATRP.

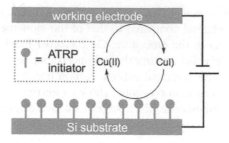

FIGURE 6.2   Schematic illustration of the setup for surface-initiated *e*ATRP.

This method can be applied to surface-initiated polymerization (SIP). Zhou and co-workers fabricated a setup to enable SIP by sandwiching an electrolyte with a cathode and an initiator-modified substrate [57,58]. In this system, Cu^I generated at the cathode diffuses to the initiator-modified substrate and catalyzes the polymerization of vinyl monomer, leading to the formation of polymer brushes (Figure 6.2). Recently, Inagi and co-workers successfully demonstrated the fabrication of gradient polymer brushes with a SIP process using a bipolar electrochemical technique [59]. A bipolar electrode [60–62], having anodic and cathodic poles in single material, can generate a potential gradient on its surface for the continuous generation of a concentration gradient of Cu^I. Because the bipolar electrolysis does not require much supporting electrolytes, nonpolar vinyl monomers such as methyl methacrylate are available for the process.

6.4.3 ELECTROCHEMICAL CROSS-LINKING REACTION OF NONCONJUGATED POLYMERS

Diffusion of electrogenerated catalysts enables the continuous growth of polymer films even when the obtained films are not conductive. Schaaf, Boulmedais, and co-workers regarded the electrogenerated Cu^I catalyst as a morphogen, a specific species to which cells respond in a concentration-dependent manner, and controlled the self-assembly of two independent polymers at the near surface of the electrode. They synthesized two kinds of homopolymers, one possessing azide groups and the other one bearing terminal alkyne groups in its side chain. In an electrolyte containing these two polymers, solvent and Cu^{II}, electrochemical reduction of Cu^{II} to Cu^I triggered the azide-alkyne cycloaddition reaction, known as click chemistry, proceeded to give cross-linked polymer films (Figure 6.3) [63]. The continuous growth of polymer films during potential dynamic electrolysis was observed by monitoring an electrochemical quartz crystal microbalance technique.

6.5 CONCLUSION AND OUTLOOK

In this chapter, the recent progress in electrochemistry for functional polymer synthesis spanning from electropolymerization to post-functionalization of conducting

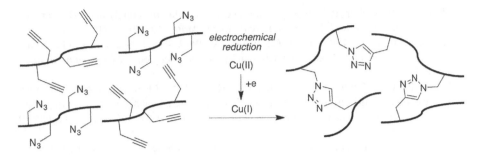

FIGURE 6.3 Formation of cross-linked polymer films on the electrode surface triggered by the cathodic reduction of Cu^{II}.

polymers based on the concept of organic electrosynthesis was described. The conducting polymers on electrodes exhibit interesting optoelectronic properties, which are potentially useful for organic electronics. The heterogeneous reaction fields of electrode/electrolyte interfaces bring advantages for preparing not only conducting and electroactive polymers but also nonconductive polymer brushes and films. Moreover, the combination of conducting polymer materials and bipolar electrochemistry has afforded the gradient electrochromic materials [64–66], composition gradient [44,67], gradient surface modification [68], and conducting polymer fiber formation [69].

REFERENCES

1. Skotheim, T.A.; Reynolds, J.R. 2007. *Handbook of Conducting Polymers,* 3rd ed. CRC Press, Boca Raton, FL.
2. Inzelt, G. 2008. *Conducting Polymers: A New Era of Electrochemistry.* Springer, Heidelberg.
3. Bhadani, S.N.; Parravano, G. 1983. Electrochemical polymerization. In *Organic Electrochemistry,* Baizer, M.M.; Lund, H. (eds.) 2nd ed., Chapter 31. Marcel Dekker, New York, 995–1029.
4. Funt, B.L. 1991. Electrochemical polymerization. In *Organic Electrochemistry,* Baizer, M.M.; Lund, H. (eds.) 3rd ed., Chapter 32. Marcel Dekker, New York, 1337–1362.
5. Heinze, J.; Frontana-Uribe, B.A.; Ludwigs, S. 2010. Electrochemistry of conducting polymers—Persistent models and new concepts. *Chem. Rev.* 110: 4724–4771.
6. Beaujuge, P.M.; Reynolds, J.R. 2010. Color control in π-conjugated organic polymers for use in electrochromic devices. *Chem. Rev.* 110: 268–310.
7. Smela, E. 2003. Conjugated polymer actuators for biomedical applications. *Adv. Mater.* 15: 481–494.
8. Ando, S.; Ueda, M. 2002. Density functional theory calculations of the local spin densities of 3-substituted thiophenes and the oligomerization mechanism of 3-methylsulfanyl thiophene. *Synth. Met.* 129: 207–213.
9. Sekiguchi, K.; Atobe, M.; Fuchigami, T. 2002. Electropolymerization of pyrrole in 1-ethyl-3-methylimidazolium trifluoromethanesulfonate room temperature ionic liquid. *Electrochem. Commun.* 4: 881–885.
10. Chen, W.; Xue, G. 2005. Low potential electrochemical syntheses of heteroaromatic conducting polymers in a novel solvent system based on trifluroborate–ethyl ether. *Prog. Polym. Sci.* 30: 783–811.

11. Huang, H.; Pickup, P.G. 1998. A donor–acceptor conducting copolymer with a very low band gap and high intrinsic conductivity. *Chem. Mater.* 10: 2212–2216.

12. Latonen, R.-M.; Kvarnström, C.; Ivaska, A. 1999. Electrochemical synthesis of a copolymer of poly(3-octylthiophene) and poly(paraphenylene). *Electrochim. Acta* 44: 1933–1943.

13. Smith, E.L.; Glidle, A.; Mortimer, R.J.; Ryder, K.S. 2007. Spectroelectrochemical responses of thin-film conducting copolymers prepared electrochemically from mixtures of 3,4-ethylenedioxythiophene and 2,2′-bithiophene. *Phys. Chem. Chem. Phys.* 9: 6098–6105.

14. Nie, G.; Qu, L.; Xu, J.; Zhang, S. 2008. Electrosyntheses and characterizations of a new soluble conducting copolymer of 5-cyanoindole and 3,4-ethylenedioxythiophene. *Electrochim. Acta* 53: 8351–8358.

15. Link, S.M.; Scheuble, M.; Goll, M.; Muks, E.; Ruff, A.; Hoffmann, A.; Richter, T.V.; Lopes Navarrete, J.T.; Ruiz Delgado, M.C.; Ludwigs, S. 2013. Electropolymerized three-dimensional randomly branched EDOT-containing copolymers. *Langmuir* 29: 15463–15473.

16. Imae, I.; Imabayashi, S.; Korai, K.; Mashima, T.; Ooyama, Y.; Komaguchi, K.; Harima, Y. 2012. Electrosynthesis and charge-transport properties of poly(3′,4′-ethylenedioxy-2,2′:5′,2″-terthiophene). *Mater. Chem. Phys.* 131: 752–756.

17. Gohier, F.; Frère, P.; Roncali, J. 2013. 3-Fluoro-4-hexylthiophene as a building block for tuning the electronic properties of conjugated polythiophenes. *J. Org. Chem.* 78: 1497–1503.

18. Kingsbrough, R.P.; Swager, T.M. 1998. Electroactivity enhancement by redox matching in cobalt salen-based conducting polymers. *Adv. Mater.* 10: 1100–1104.

19. Ruhlmann, L.; Schulz, A.; Giraudeau, A.; Messerschmidt, C.; Fuhrhop, J.-H. 1999. A polycationic zinc-5,15-dichlorooctaethylporphyrinate-viologen wire. *J. Am. Chem. Soc.* 121: 6664–6667.

20. Saito, N.; Kanbara, T.; Nakamura, Y.; Yamamoto, T.; Kubota, K. 1994. Electrochemical and chemical preparation of linear π-conjugated poly(quinoline-2,6-diyl) using nickel complexes and electrochemical properties of the polymer. *Macromolecules* 27: 756–761.

21. Utley, J.H.P.; Gruberb, J. 2002. Electrochemical synthesis of poly(*p*-xylylenes) (PPXs) and poly(*p*-phenylenevinylenes) (PPVs) and the study of xylylene (quinodimethane) intermediates; an underrated approach. *J. Mater. Chem.* 12: 1613–1624.

22. Miller, R.D.; Michl, J. 1989. Polysilane high polymers. *Chem. Rev.* 89: 1359–1410.

23. Shono, T.; Kashimura, S.; Ishifune, M.; Nishida, R. 1990. Electroreductive formation of polysilanes. *J. Chem. Soc., Chem. Commun.* 17: 1160–1161.

24. Kashimura, S.; Ishifune, M.; Yamashita, N.; Bu, H.-B.; Takebayashi, M.; Kitajima, S.; Yoshiwara, D.; Kataoka, Y.; Nishida, R.; Kawasaki, S.-I.; Murase, H.; Shono, T. 1999. Electroreductive synthesis of polysilanes, polygermanes, and related polymers with magnesium electrodes. *J. Org. Chem.* 64: 6615–6621.

25. Shono, T.; Kashimura, S.; Murase, H. 1992. Electroreductive synthesis of polygermane and germane–silane copolymer. *J. Chem. Soc., Chem. Commun.* 12: 896–897.

26. Inagi, S.; Fuchigami, T. 2014. Electrochemical post-functionalization of conducting polymers. *Macromol. Rapid Commun.* 35: 854–867.

27. Qi, Z.; Pickup, P.G. 1992. Reactivation of poly(3-methylthiophene) following overoxidation in the presence of chloride. *J. Chem. Soc., Chem. Comm.* 22: 1675–1676.

28. Qi, Z.; Pickup, P.G. 1993. X-ray emission analysis of thin poly(3-methylthiophene) and poly{(3-methylthiophene)-co-[1-methyl-3-(pyrrol-1-ylmethyl)pyridinium]} films. Composition, oxidation level, and overoxidation. *Anal. Chem.* 65: 696–703.

29. Qi, Z.; Rees, N.G.; Pickup, P.G. 1996. Electrochemically induced substitution of polythiophenes and polypyrrole. *Chem. Mater.* 8: 701–707.

30. Soudan, P.; Lucas, P.; Breau, L.; Bélanger, D. 2000. Electrochemical modification of poly(3-(4-fluorophenyl)thiophene). *Langmuir* 16: 4362–4366.

31. Li, Y.; Kamata, K.; Asaoka, S.; Yamagishi, T.; Iyoda, T. 2003. Efficient anodic pyridination of poly(3-hexylthiophene) toward post-functionalization of conjugated polymers. *Org. Biomol. Chem.* 1: 1779–1784.

32. Fabre, B.; Simonet, J. 1996. Post-polymerization electrochemical functionalization of a conducting polymer: anodic cyanation of poly(*p*-dimethoxybenzene). *J. Electroanal. Chem.* 416: 187–189.

33. Fabre, B.; Kanoufi, F.; Simonet, J. 1997. Electrochemical and XPS investigations of the anodic substitution of an electronic conducting polymer. Cyanation of poly-[(1,4-dimethoxybenzene)-*co*-(3-methylthiophene)]. *J. Electroanal. Chem.* 434: 225–234.

34. Inagi, S.; Hayashi, S.; Hosaka, K.; Fuchigami, T. 2009. Facile functionalization of a thiophene–fluorene alternating copolymer via electrochemical polymer reaction. *Macromolecules* 42: 3881–3883.

35. Inagi, S.; Hosaka, K.; Hayashi, S.; Fuchigami, T. 2010. Solid-phase halogenation of a conducting polymer film via electrochemical polymer reaction. *J. Electrochem. Soc.* 157: E88–E91.

36. Hayashi, S.; Inagi, S.; Hosaka, K.; Fuchigami, K. 2009. Post-functionalization of poly (3-hexylthiophene) via anodic chlorination. *Synth. Met.* 159: 1792–1795.

37. Hayashi, S.; Inagi, S.; Fuchigami, T. 2011. Efficient electrochemical polymer halogenation using a thin-layered cell. *Polym. Chem.* 2: 1632–1637.

38. Fuchigami, T.; Inagi, S. 2011. Selective electrochemical fluorination of organic molecules and macromolecules in ionic liquids. *Chem. Commun.* 47: 10211–10223.

39. Inagi, S.; Hayashi, S.; Fuchigami, T. 2009. Electrochemical polymer reaction: selective fluorination of a poly(fluorene) derivative. *Chem. Commun.* 45: 1718–1720.

40. Hayashi, S.; Inagi, S.; Fuchigami, T. 2009. Synthesis of 9-substituted fluorene copolymers via chemical and electrochemical polymer reaction and their optoelectronic properties. *Macromolecules* 42: 3755–3760.

41. Hayashi, S.; Inagi, S.; Fuchigami, T. 2010. Electrochemical modification of a fluorene-carbazole alternating copolymer toward a novel donor-acceptor type conjugated polymer. *Electrochemistry* 78: 114–117.

42. Inagi, S.; Koseki, K.; Hayashi, S.; Fuchigami, T. 2010. Electrochemical tuning of the optoelectronic properties of a fluorene-based conjugated polymer. *Langmuir* 26: 18631–18633.

43. Baizer, M.M. 1991. Paired electrosynthesis. In *Organic Electrochemistry*, Baizer, M.M.; Lund, H. (eds.) 3rd ed., Chapter 35, 1421–1438, Marcel Dekker, New York.

44. Inagi, S.; Nagai, H.; Tomita, I.; Fuchigami, T. 2013. Parallel polymer reactions of a polyfluorene derivative by electrochemical oxidation and reduction. *Angew. Chem. Int. Ed.* 52: 6616–6619.

45. Han, C.-C.; Jeng, R.-C. 1997. Concurrent reduction and modification of polyaniline emeraldine base with pyrrolidine and other nucleophiles. *Chem. Commun.* 553–554.

46. Han, C.-C.; Hong, S.-P.; Yang, K.-F.; Bai, M.-Y.; Lu, C.-H.; Huang, C.-S. 2001. Highly conductive new aniline copolymers containing butylthio substituent. *Macromolecules*, 34: 587–591.

47. Han, C.-C.; Chen, H.-Y. 2007. Highly conductive and electroactive fluorine-functionalized polyanilines. *Macromolecules* 40: 8969–8973.

48. Song, C.; Swager, T.M. 2009. Conducting polymers containing *peri*-xanthenoxanthenes via oxidative cyclization of binaphthols. *Macromolecules* 42: 1472–1475.

49. Abruña, H.D.; Denisevich, P.; Umana, M.; Meyer, T.J.; Murray, R.W. 1981. Rectifying interfaces using two-layer films of electrochemically polymerized vinylpyridine and vinylbipyridine complexes of ruthenium and iron on electrodes. *J. Am. Chem. Soc.* 103: 1–5.
50. Zhong, Y.-W.; Yao, C.-J.; Nie, H.-J. 2013. Electropolymerized films of vinyl-substituted polypyridine complexes: synthesis, characterization, and applications. *Coord. Chem. Rev.* 257: 1357–1372.
51. Nie, H.-J.; Shao, J.-Y.; Wu, J.; Yao, J.; Zhong, Y.-W. 2012. Synthesis and reductive electropolymerization of metal complexes with 5,5′-divinyl-2,2′-bipyridine. *Organometallics* 31: 6952–6959.
52. Matyjaszewski, K.; Xia, J.H. 2001. Atom transfer radical polymerization. *Chem. Rev.* 101: 2921–2990.
53. Kamigaito, M.; Ando, T.; Sawamoto, M. 2001. Metal-catalyzed living radical polymerization. *Chem. Rev.* 101: 3689–36745.
54. Magenau, A.J.D.; Strandwitz, N.C.; Gennaro, A.; Matyjaszewski, K. 2011. Electrochemically mediated atom transfer radical polymerization. *Science* 332: 81–84.
55. Bortolamei, N.; Isse, A.A.; Magenau, A.J.D.; Gennaro, A.; Matyjaszewski, K. 2011. Controlled aqueous atom transfer radical polymerization with electrochemical generation of the active catalyst. *Angew. Chem. Int. Ed.* 50: 11391–11394.
56. Magenau, A.J.D.; Bortolamei, N.; Frick, E.; Park, S.; Gennaro, A.; Matyjaszewski, K. 2013. Investigation of electrochemically mediated atom transfer radical polymerization. *Macromolecules* 46: 4346–4353.
57. Li, B.; Yu, B.; Huck, W.T.S.; Zhou, F.; Liu, W. 2012. Electrochemically induced surface-initiated atom-transfer radical polymerization. *Angew. Chem. Int. Ed.* 51: 5092–5095.
58. Li, B.; Yu, B.; Huck, W.T.S.; Liu, W.; Zhou, F. 2013. Electrochemically mediated atom transfer radical polymerization on nonconducting substrates: controlled brush growth through catalyst diffusion. *J. Am. Chem. Soc.* 125: 1708–1710.
59. Shida, N.; Koizumi, Y.; Nishiyama, H.; Tomita, I.; Inagi, S. 2015. Electrochemically mediated atom transfer radical polymerization from a substrate surface manipulated by bipolar electrolysis: fabrication of gradient and patterned polymer brushes. *Angew. Chem. Int. Ed.* 54: 3922–3926.
60. Loget, G.; Zigah, D.; Bouffier, L.; Sojic, N.; Kuhn, A. 2013. Bipolar electrochemistry: from materials science to motion and beyond. *Acc. Chem. Res.* 46: 2513–2523.
61. Fosdick, S.E.; Knust, K.N.; Scida, K.; Crooks, R.M. 2013. Bipolar electrochemistry. *Angew. Chem. Int. Ed.* 52: 10438–10456.
62. Inagi, S. 2016. Fabrication of gradient polymer surfaces using bipolar electrochemistry. *Polym. J.* 48: 39–44.
63. Rydzek, G.; Jierry, L.; Parat, A.; Thomann, J.-S.; Voegel, J.-C.; Senger, B.; Hemmerlé, J.; Ponche, A.; Frisch, B.; Schaaf, P.; Boulmedais, F. 2011. Electrochemically triggered assembly of films: a one-pot morphogen-driven buildup. *Angew. Chem. Int. Ed.* 50: 4374–4377.
64. Inagi, S.; Ishiguro, Y.; Atobe, M.; Fuchigami, T. 2010. Bipolar patterning of conducting polymers by electrochemical doping and reaction. *Angew. Chem. Int. Ed.* 49: 10136–10139.
65. Ishiguro, Y.; Inagi, S.; Fuchigami, T. 2011. Gradient doping of conducting polymer films by means of bipolar electrochemistry. *Langmuir* 27: 7158–7162.
66. Inagi, S.; Ishiguro, Y.; Shida, N.; Fuchigami, T. 2012. Measurements of potential on and current through bipolar electrode in U-type electrolytic cell with a shielding wall. *J. Electrochem. Soc.* 159: G146–G150.

67. Ishiguro, Y.; Inagi, S.; Fuchigami, T. 2012. Site-controlled application of electric potential on a conducting polymer "Canvas." *J. Am. Chem. Soc.* 134: 4034–4036.
68. Shida, N.; Ishiguro, Y.; Atobe, M.; Fuchigami, T.; Inagi, S. 2012. Electro-click modification of conducting polymer surface using Cu(I) species generated on a bipolar electrode in a gradient manner. *ACS Macro. Lett.* 1: 656–659.
69. Koizumi, Y.; Shida, N.; Ohira, M.; Nishiyama, H.; Tomita, I.; Inagi S. 2016. Electropolymerization on wireless electrodes towards conducting polymer microfibre networks. *Nat. Commun.* 7: 10404.

7 Biphasic and Emulsion Electroorganic Syntheses

Sunyhik Ahn and Frank Marken
University of Bath

CONTENTS

7.1 Introduction to Emulsion and Biphasic Electrosynthesis 149
7.2 Surfactant-Stabilized Biphasic Electrolysis 154
7.3 Surfactant-Free Biphasic Electrolysis 155
7.4 Triple-Phase Boundary Electrolysis 157
7.5 Conclusion and Outlook 160
References 162

7.1 INTRODUCTION TO EMULSION AND BIPHASIC ELECTROSYNTHESIS

The importance of aqueous emulsions and biphasic methods in the development of "green electrosynthesis" has been emphasized in a review by Frontana-Uribe et al. [1]. The "greenness" of these processes is clearly linked to economic benefits (use of water), which make biphasic methods attractive in industrial-scale electrolytic processes. The benefits of using emulsion-based systems for syntheses can be demonstrated by redox emulsion polymerization reactions of polyaniline (PANI) developed by Monsanto company (see Figure 7.1 [2]). Here, a higher concentration of aniline monomers is concentrated in the organic phase component of the microemulsions. Use of an aqueous chemical oxidant $(NH_4)_2S_2O_8$ allows the polymerization reaction to proceed effectively, while the reactant and product remain in separate phases from oxidant and waste products such as $(NH_4)_2SO_4$. Microemulsions offer a high interfacial surface area between aqueous and organic phases for an enhanced rate of reaction. As this reaction progresses, the microemulsion flocculates to give PANI product phase-separated with minimal contamination. This is significant as additional post-synthetic processing steps involving purification and phase separation are undesirable in terms of cost and atom economy. The microemulsion synthesis method offers both greener reaction conditions and better separation of products.

The atom economy of emulsion-based synthesis can be further enhanced by use of an applied potential for direct PANI oxidation or for the recycling of the redox mediator peroxodisulfate. Many other electrosyntheses can be performed in part or in total under emulsion conditions. The important Monsanto process for the production of nylon [3,4] offers an example, for which gap cells were operated with aqueous

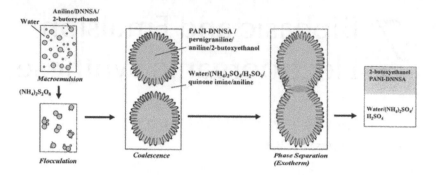

FIGURE 7.1 Schematic drawing of an electroorganic emulsion redox reaction with a chemical oxidant $(NH_4)_2S_2O_8$, dinonylnaphthalenesulfonic acid (DNNSA), and 2-butoxyethanol surfactant and cosurfactant. PANI product remains in the organic phase (in pink), while the reactant and by-products of the reaction remain in the aqueous phase. (From Kinlen, P.J. et al., *Macromolecules*, 31, 1735, 1998. With permission. Copyright 1998 American Chemical Society.)

emulsion media flowing through to electroreductively couple two equivalents of acrylonitrile to give adiponitrile. Further authoritative review articles on emulsion electrolysis techniques [5] and applications [6] have appeared. The general benefits and applications of surfactants applied in organic syntheses have been emphasized [7].

Electrolysis reactions when carried out in conventional electrolysis systems with aqueous electrolyte media generally suffer from the low solubility of the organic reactants (dispersed or in emulsion) in the aqueous phase. Without further measures, this may lead to very low currents and therefore ineffective electrolysis. A significant improvement is achieved when introducing aqueous-soluble mediator redox systems such as $Cr^{2+/3+}$ or $Ce^{3+/4+}$ (see Figure 7.2) to more effectively transfer electrons from the reagent to the electrode. A reactor system can be based on electrolytic regeneration and a separate reactor for the emulsion reaction (Figure 7.2).

There are now many more examples of redox mediator systems [8], often with specific catalytic activity or selectivity. Further improvements in reaction rate have been introduced in the form of phase transfer catalysts [9–11] that operate at the organic\aqueous interface and allow redox mediator molecules to enter the organic phase. Phase transfer catalysts form a complex (or ion pair) with the oxidant/reductant and thereby transfer these to the surface or the bulk of the organic phase to greatly accelerate the rate of reaction. An example of anthracene electrooxidation to anthraquinone (with anthracene as solid in a slurry electrolyte) with the phase transfer catalyst dodecyl-benzene-sulfonate [12] is shown in Figure 7.3. In this work, Chou et al. investigate the mechanism and optimize the conditions for this multielectron slurry oxidation to give high yields catalyzed by the amphiphilic phase transfer catalyst dodecyl-benzene-sulfonate that allows the Mn(III) cation to effectively interact with the anthracene starting material. The overall type of this reaction appears to be solid to solid.

There are many examples of interesting emulsion-based electroorganic transformation as documented in Table 7.1. This includes oxidation reactions, chlorination, bromination, cyanation, acyloxylation, reduction, nitration, epoxidation, and coupling reactions.

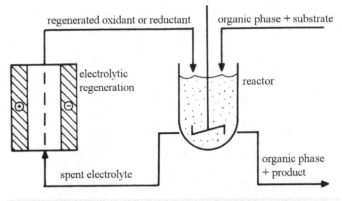

Redox Couple	Standard Potential (V)	
$[Co(CN)_6]^{4-}/[Co(CN)_6]^{3-}$	-0.83	Reduction
Cr^{2+}/Cr^{3+}	-0.41	
Ti^{2+}/Ti^{3+}	-0.37	
Sn^{2+}/Sn^{4+}	$+0.15$	
$[Fe(CN)_6]^{4-}/[Fe(CN)_6]^{3-}$	$+0.36$	
U^{4+}/U^{6+}	$+0.40$	
MnO_4^{2-}/MnO_4^{-}	$+0.54$	
$VO^{2+}/V(OH)_4^{+}$	$+1.00$	
$3\,Br^{-}/Br_3^{-}$	$+1.05$	
$Cr^{3+}/Cr_2O_7^{2-}$	$+1.36$	
Ce^{3+}/Ce^{4+}	$+1.44$	
Mn^{2+}/Mn^{3+}	$+1.51$	
Mn^{2+}/MnO_4^{-}	$+1.52$	
Pb^{2+}/Pb^{4+}	$+1.69$	
Co^{2+}/Co^{3+}	$+1.84$	Oxidation

FIGURE 7.2 Schematic drawing of an electroorganic emulsion reactor with a redox mediator regeneration cycle. Also shown is a list of typical aqueous redox mediators and their standard potentials. (From Feess, H. and Wendt, H., *Techniques of Electro-Organic Synthesis*, John Wiley & Sons, New York, 1982. With permission.)

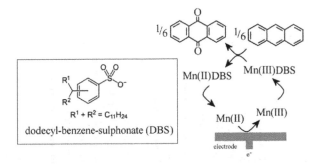

FIGURE 7.3 Molecular structure of dodecyl-benzene-sulfonate and schematic drawing of the slurry oxidation of anthracene to anthraquinone. (From Chou, T.C. and Cheng, C.H., *J. Appl. Electrochem.*, 22, 743, 1992. With permission.)

TABLE 7.1

Summary of Representative Emulsion Electroorganic Transformations

Year	Reaction	Conditions	Electrode	Solvents	Yields	Ref.
1977	Alcohol oxidation	Hypobromite (oxidized aqueous bromide)	Platinum	Amyl acetate/water emulsion with 2% tetrabutylammonium bisulfate	60%–70% after 2 F mol⁻¹ passed (quantified by gas chromatography)	[11]
1983	Chlorination of naphthalenes	$ZnCl_4[Bu_4N]_2$	Platinum	Methylene chloride/water emulsion with tetrabutylammonium	92% mono-chloro product after 2.33 F mol⁻¹ quantified by GC	[13,14]
1982	Cyanation of naphthalene	$Bu_4N(CN)$	Platinum	Methylene chloride/water emulsion with tetrabutylammonium	79% mono-product after 2.5 F mol⁻¹ quantified by GC	[15]
1982	Acyloxylation of dimethoxy-benzenes	$Bu_4N(Oac)$	Platinum	Methylene chloride/water emulsion with tetrabutylammonium	87% mono-product after 2.33 F mol⁻¹ quantified by GC	[16]
1984	Cyanation of dimethoxy-benzenes	$Bu_4N(CN)$	Platinum	Dichloromethane/water emulsion with tetrabutylammonium	81% mono-product after 2 F mol⁻¹ quantified by IR, NMR, and mass spectroscopy	[17]
1992	Reduction of 2-ethyl-anthraquinone	Direct	Lead-plated glassy carbon	Ethylbenzene/water		[18]
1996	Benzyl alcohol oxidation	NiOOH	Nickel foam	Petroleum ether/water	40% in 4 h quantified by GC	[19]
1997	Reduction of nitrobenzenes	$(C_5H_5)_2TiOH$	Graphite	Dichloromethane, toluene/water	Products identified by TLC, IR. and NMR	[20]
1999	Nitration of naphthalene	NO_2^- (from $NaNO_2$)	Platinum	Brij 35 (polyoxyethylene-23-dodecanol)/water	Products identified with HPLC	[21]
1999	Reduction of 2-ethyl-9,10-anthraquinone		Glassy carbon	Tributylphosphate + diethylbenzene/aqueous phase		[22]
2001	Olefin epoxidation	Mn-salen complex	Platinum	CH_2Cl_2/water	87% over 16h; products identified with HPLC	[23]

(Continued)

TABLE 7.1 (Continued)
Summary of Representative Emulsion Electroorganic Transformations

Year	Reaction	Conditions	Electrode	Solvents	Yields	Ref.
2002	Synthesis of dibenzonaphthyridine derivatives	$(C_5H_5)_2Ti$	Graphite	CH_2Cl_2/water	Product identified with NMR	[24]
2006	Toluene bromination	BrOH	Platinum	CH_2Cl_2/water	Product identified with HPLC	[25]
2008	Homolytic/heterolytic olefin coupling	Benzyl bromide (not catalytic)	Graphite	Hexane+butanol/water	Product identified with HPLC and NMR	[26]
2011	Olefine reduction		Carbon nanofibers	Acetonitrile/water	Product identified by NMR	[27]
2012	Benzyl alcohol oxidation		Carbon	Chloroform/water	95% yield after 2 F mol^{-1}. Products identified with HPLC	[28]
2012	Olefin epoxidation	Sodium hypochlorite	Glassy carbon	Ionic liquid (BMImBF$_4$)/water	90% yield 2 h. Products identified with HPLC	[29]
2013	Benzylic C–H bond functionalization	Bromide ion/TEMPO	Glassy carbon	CH_2Cl_2/water	70% after 7.5 F mol^{-1}. Product quantified with GC and NMR	[30]

GC, gas chromatography; HPLC, high-performance liquid chromatography; IR, infrared; NMR, nuclear magnetic resonance spectroscopy; TLC, thin layer chromatography.

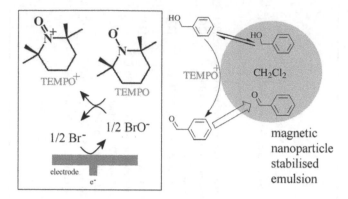

FIGURE 7.4 Redox mediator chain with TEMPO as active redox catalyst. Schematic drawing of the benzylalcohol to benzaldehyde transformation under emulsion conditions. (From Anelli, P.L. et al., *J. Org. Chem.*, 52, 2559, 1987. With permission.)

Recently, Jing and coworkers [31] demonstrated intensification of emulsion electrosynthesis of benzaldehyde from benzylalcohol based on the Montanari reaction [32] (see Figure 7.4). This process uses the aqueous bromide/hypobromide redox mediator, which oxidizes the redox catalyst TEMPO to the active form $TEMPO^+$. The emulsion droplets are stabilized with magnetic nanoparticles (for easy separation) and effects of mass transport and temperature on reaction rate are evaluated. A further development was reported by Tang based on the immobilization of the TEMPO catalyst directly within the nanoparticulate surfactant [33].

7.2 SURFACTANT-STABILIZED BIPHASIC ELECTROLYSIS

Rusling and coworkers [34,35] have pointed out the benefits of surfactant-stabilized microemulsion in electrolytic transformations. Although adding surfactant can lead to obvious complexity in the workup of products, there are many cases where this is not a problem (e.g., for volatile products), and where good yields can be achieved as well as electrolyte recycling. With the help of surfactants and cop-surfactants, "microemulsion" conditions can be achieved with emulsion droplets with diameter of 100 nm or less. These microemulsions are optically clear, and they provide a highly extended liquid–liquid interfacial region for phase-transfer and redox processes to proceed. An example of a microemulsion electroorganic transformation based on a catalytic metal complex in organic electrosynthesis [36] for the natural redox mediator and catalyst vitamin B_{12} [37] is shown in Figure 7.5. Rusling and coworkers employed didodecyl-dimethyl-ammonium bromide surfactant and dimethyl-formamide-water microemulsion conditions. The process was optimized to give close to 100% yield in dibenzyl without toluene side products.

Recently, Compton and coworkers reinvestigated the vitamin B_{12} microemulsion catalysis at the single microdroplet level [38]. Dehalogenation of trichloroethylene during "soft nanoparticle impacts" was studied quantitatively, and the catalytic turnover number was evaluated by investigating the current spikes generated when organic microdroplets interact individually with the electrode.

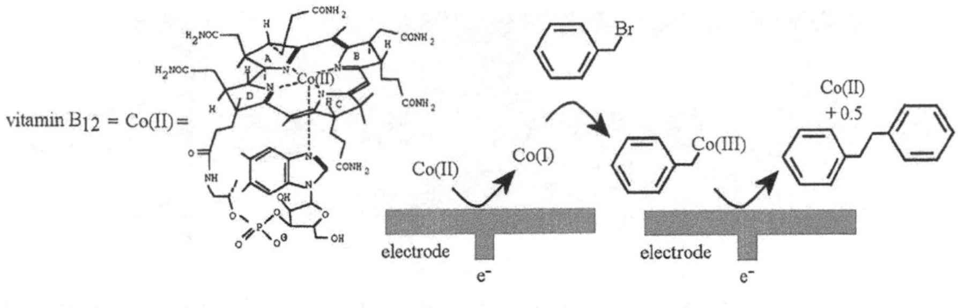

FIGURE 7.5 Molecular structure of vitamin B_{12} and schematic description of the consecutive two-electron reduction leading to biphenyl. (Adapted from Zhou, D.L. et al. *Langmuir*, 12, 3067, 1996. With permission.)

Nano-materials (such as nanoparticulate or nanotubular carbon) often show amphiphilic properties as they lower surface tension at the interface of organic and aqueous liquid phases. As a result, these nano-materials get attracted into the liquid|liquid interfacial region (forming Pickering emulsion systems [39]) to stabilize and modify the interface. In particular, graphene or nano-carbon materials [40] could be very useful in emulsion electrosyntheses as surfactant as well as redox catalysts.

7.3 SURFACTANT-FREE BIPHASIC ELECTROLYSIS

In order to comply with the pillars of green chemistry [41], ideally emulsion electrosynthesis should be carried out without additives such as surfactants, for example, aided by mechanically induced emulsification. In a review on biphasic electrolysis methods, ultrasound emulsification has been suggested as a way to avoid surfactants [42]. The formation of emulsion droplets under ultrasound conditions relies on high shear forces and forceful mixing. As soon as the source of the ultrasound is switched off, the emulsion will separate out to allow effective product recovery.

The interaction of microdroplet deposits at electrode surfaces is sensitive to the presence of surfactant and has been investigated for a range of redox systems [43]. For immobilized microdroplets, the energy requirements for interfacial ion transfer have been shown to be linked to the Gibbs energy of transfer and to the electron transfer [44]. Single microdroplet collision experiments demonstrating the rate and mechanism of organic microdroplet reactions at microelectrodes have been observed, for example, by Bard and coworkers [45]. One particular benefit in biphasic electrolysis is the possibility to work in a single compartment without a separating membrane and with distinctly different anode and cathode reactions [46] and no interference.

In addition to power ultrasound agitation, ultra-turrax agitation, which involves rotation of a dense arrangement of small shearing blades, has been shown to allow acetonitrile-aqueous emulsions to be formed in situ and employed in electrolysis [47]. Figure 7.6 shows the single-compartment electrolysis cell with two separate organic and aqueous phases. During operation (Figure 7.6B), the two phases are emulsified and after completion of the electrolysis, the two phases separate out again.

FIGURE 7.6   Electrolysis cell for ultra-turrax emulsification (A) before and (B) during operation. (C) Schematic drawing of components (i) ultra-turrax insert, (ii) working electrode, (iii) counter electrode, and (iv) reference electrode. (From Watkins, J.D. et al., *Electrochim. Acta*, 55, 8808, 2010. With permission.)

The redox reaction at the location where *n*-butylferrocene is oxidized to *n*-butylferricinium requires the balance of charges, and therefore the oxidation in the organic phase is associated with the transfer of an anion from the aqueous into the organic phase. This can be observed directly by comparing voltammetric signals for different types of electrolyte. The presence of KPF_6 in the aqueous phase (Figure 7.7C) causes a shift of the oxidation to more negative potentials due to the lower transfer potential for PF_6^-. The drawing in Figure 7.7B suggests that the redox reaction occurs at the triple-phase boundary electrode|organic phase|aqueous phase, which demonstrates the importance of high shear forces producing very small microdroplets and promotes the fast exchange of organic phase at the electrode surface and in solution. This schematic drawing also suggests that the type of electrode surface (roughness, hydrophobicity, surface tension) may be crucial in making this type of electrolysis effective.

Power ultrasound offers another convenient high-shear methodology to work with intrinsically unstable emulsions that are maintained by the energy from the sonic field [48]. Davies et al. demonstrated the surfactant-free C–C coupling process [49] under power ultrasound electrolysis conditions. The oxidation of amines and the

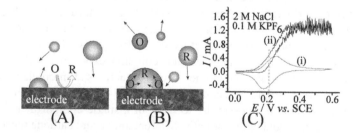

FIGURE 7.7   Schematic drawing of the reaction conditions with organic microdroplets (containing yellow *n*-butylferrocene) interacting with the surface and reacting (A) via dissolution and electron transfer in the aqueous phase or (B) at the triple-phase boundary organic|aqueous|electrode. (C) Typical voltammogram for the oxidation of *n*-butylferrocene in acetonitrile (i) without and (ii) with ultra-turrax agitation. The presence of KPF_6 defines the oxidation potential due to transfer of PF_6^- from the aqueous into the organic phase during oxidation. (From Watkins, J.D. et al., *Electrochim. Acta,* 55, 8808, 2010. With permission.)

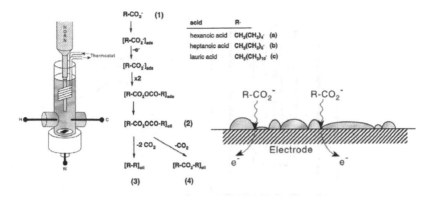

FIGURE 7.8 Schematic drawing of the experimental setup for the biphasic sono-Kolbe electrolysis. The reaction scheme shows the mechanistic pathway with the dimer (3), the main Kolbe product. The schematic drawing shows conditions at the electrode surface during electrolysis. (From Wadhawan, J.D. et al., *J. Electroanal. Chem.*, 507, 135, 2001. With permission.)

electroreduction of olefins in sono-emulsion systems were investigated by Atobe et al. [50]. Tandem ultrasound methods have been suggested for electropolymerization applications [51]. The simultaneous application of two different frequencies of ultrasound causes a much improved microdroplet control and much lower droplet sizes.

Power ultrasound-assisted emulsion electrolysis has been demonstrated by Wadhawan et al. for the case of the biphasic Kolbe electrolysis processes [52] (see Figure 7.8). On both platinum- and boron-doped diamond electrodes, similar processes were observed with the Kolbe dimer (see (3) in Figure 7.8) being produced as the main product. Note that the counter electrode is not separated and that with a careful choice of size and positioning of the platinum counter electrode, unwanted cathodic reactions can be suppressed. It is also interesting to note that a boron-doped diamond working electrode under these conditions can be used instead of platinum.

7.4 TRIPLE-PHASE BOUNDARY ELECTROLYSIS

Triple-phase boundary reaction conditions [53] can be formed under carefully controlled microfluidic channel flow conditions. Flow reactors with microfluidic biphasic flow can be employed to establish/confine the triple-phase boundary reaction environment on the working electrode surface for a range of electrolytic reactions [54,55]. The development of microfluidic devices for electrosynthesis has been possible only over the past decade, but the concept of the triple-phase boundary reaction zone for biphasic reactions has been known already in the 1950s. Pioneering work by Gosh et al. [56–58] demonstrated that electrodes (here the anode) can be designed to allow reaction of water-insoluble reagents dissolved in nonpolar solvents (see Figure 7.9).

Similar to the case of the porous anode, there are many ways to design extended liquid–liquid interfaces for enhanced electrochemical reactivity. Nano-carbon-based electrodes can be employed to separate organic and aqueous liquid phases [59].

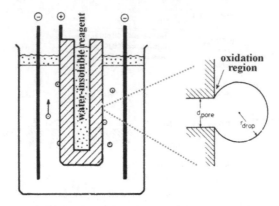

FIGURE 7.9 Schematic drawing of the electrolysis cell with highly porous cylindrical anode and conventional cathode. A water-insoluble reagent is dissolved in the organic phase which permeates through small pores to give microdroplets with a high triple-phase boundary reaction zone. (From Feess, H. and Wendt, H., *Techniques of Electro-Organic Synthesis*, John Wiley & Sons, New York, 1982. With permission.)

This type of reactor has been employed for the reduction of imines [60] (see Figures 7.10 and 7.11).

Figure 7.10 shows electron optical micrographs for "bucky-paper" electrodes with typically 50 nm diameter carbon fibers. This carbon material is amphiphilic and attracted into the liquid–liquid interface between aqueous and organic phases (see Figure 7.10D). The microelectrolysis cell in Figure 7.11A was employed with organic phase inside (with micro-propeller agitation) to form an effective triple-phase boundary reaction zone inside of the bucky paper. Figure 7.11E shows the effect of *n*-butylferrocene concentration on the voltammetric signal. Electrolysis experiments were performed for the reduction of aldehydes and of imines (formed in situ). Figure 7.12 shows the reaction scheme and two types of reaction products that were formed depending on proton availability from the aqueous phase. Both one-electron transformation and dimerization (for lower proton availability) and two-electron transformation (for higher proton availability) were reported.

In addition to carbon nanotube materials, carbon nanoparticles have been suggested as a way to extend the triple-phase boundary region where liquid–liquid ion transfer occurs [61]. Instead of modifying macroscopic electrodes, it is possible to design microscale flow devices with biphasic reaction conditions. Electrochemical devices can be miniaturized and operated under microfluidic conditions [62]. There is considerable interest in electroorganic syntheses performed under microfluidic conditions [63,64]. Single microchannel systems can be "numbered up" to stacks [65] to allow high-volume product formation. Microfluidic channel systems have been designed for "cogeneration" reactions [66] where both anode and cathode are "coupled" to perform each useful electrosynthetic transformation.

Microfluidic systems [67] have been developed for biphasic electrolysis conditions either based on microdroplet flow [68] or based on a stable liquid|liquid phase boundary in contact to the electrode surface [54,55,69]. A stable liquid–liquid interface is formed in contact to the electrode surface (see Figure 7.12). Experimental

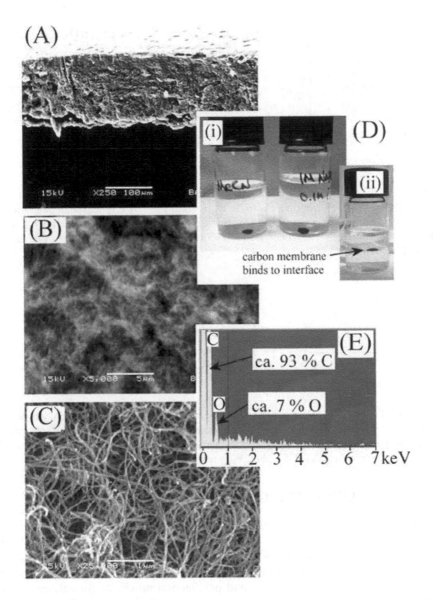

FIGURE 7.10 (A–C) Scanning electron micrographs of the "bucky-paper" electrode. (D) Photographic evidence for this nano-carbon electrode sinking to the ground in aqueous and in organic media but being immobilized at the liquid–liquid interface under biphasic conditions. (From Watkins, J.D. et al., *Tetrahedron Lett.*, 53, 3357, 2012. With permission.)

evidence suggests that this configuration allows processes to proceed without the need for added electrolyte in the organic phase. The redox reaction occurs at the electrode|organic|aqueous triple–phase boundary, and the flow rate dependence of the limiting current has been investigated [70,71] (Figure 7.13).

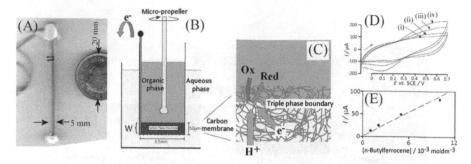

FIGURE 7.11 (A) Photograph and (B, C) schematic drawing of the microelectrolysis cell with bucky-paper working electrode at the bottom. (D) Cyclic voltammograms for *n*-butylferrocene oxidation in the organic phase as a function of concentration (i)–(iv) and (E) plot of the current versus concentration. (From Watkins, J.D. et al., *Tetrahedron Lett.*, 53, 3357, 2012. With permission.)

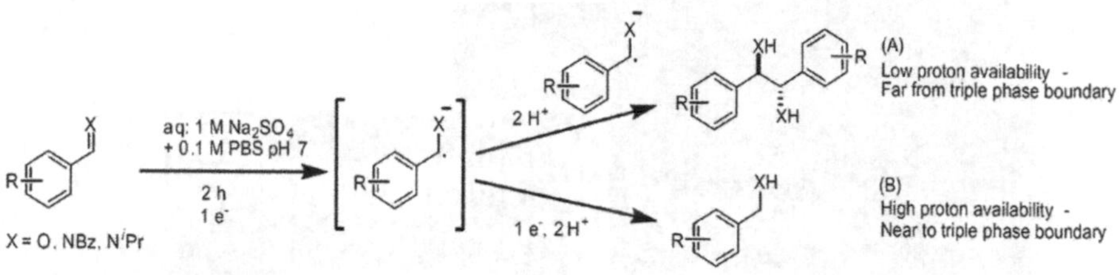

FIGURE 7.12 Reaction scheme for biphasic electroorganic transformation of aldehydes and imines. The reactions were performed with a porous bucky-paper electrode in the presence of different types of aqueous electrolyte. PBS, phosphate buffer solution. (From Watkins, J.D. et al., *Tetrahedron Lett.*, 53, 3357, 2012. With permission.)

7.5 CONCLUSION AND OUTLOOK

Biphasic and emulsion electrosyntheses provide two fundamental benefits over conventional single-phase synthesis systems based in water: (i) enhanced solubility hence wider accessibility of a range of organic substrates and reagents, and (ii) phase separation of products from mediators/reagents, which may remove the need for post-synthetic separation and purification processes for enhanced atom economy and cost efficiency. However, there are challenges with utilizing such systems such as low throughput due to low electrical conductivity of organic media and low collision frequency between reactant and interfacial oxidant (chemical or electrochemical).

This chapter provides an overview of novel methodologies developed to overcome the limitations of microemulsion or biphasic electrosyntheses, such as use of electrochemical mediators; need for enhancing mass-transport conditions, for example, by ultrasonication and mechanical agitation; and enhancing the extent of the liquid–liquid phase boundary, for example, based on the use of microemulsions or amphiphilic electrodes with a large surface area such as bucky-paper electrodes.

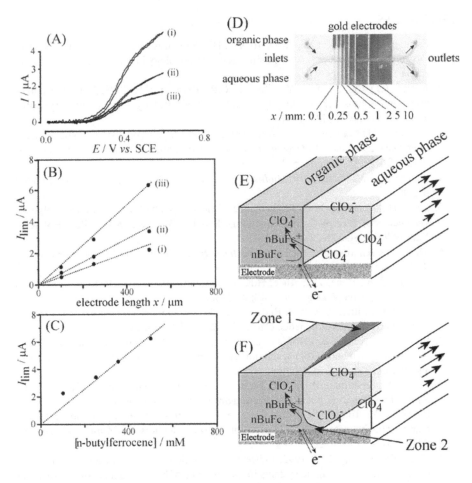

FIGURE 7.13 Biphasic voltammetry for the oxidation of *n*-butylferrocene in the flowing organic phase in contact to a flowing aqueous electrolyte phase. (A) Typical steady-state voltammograms obtained under flow conditions with plots for limiting current versus electrode length (B) and limiting current versus *n*-butylferrocene concentration (C). (D) Photograph of the biphasic flow design with gold working electrodes within the channel. (E) Schematic drawing of the two-phase flow with electron transfer and ion transfer in the triple-phase boundary region. (F) Schematic drawing of the interdiffusion region (zone 1) causing a current increase at lower flow rates. (From MacDonald, S.M. et al., *Electrochem. Commun.*, 9, 2105, 2007. With permission.)

There remain important questions to be asked addressing the reaction mechanisms under biphasic conditions. The high substrate concentration in the biphasic reaction zone is advantageous when performing redox polymerization reactions but can be detrimental in reactions where competitive monomolecular and bimolecular pathways exist. The recently emerging methodology for the study of biphasic reaction mechanisms under single microdroplet conditions could provide new insights and offer new tools for reaction control.

REFERENCES

1. Frontana-Uribe, B.A.; Little, R.D.; Ibanez, J.G.; Palma, A.; Vasquez-Medrano, R. 2010. Organic electrosynthesis: A promising green methodology in organic chemistry. *Green Chem.* 12:2099–2119.
2. Kinlen, P.J.; Liu, J.; Ding, Y.; Graham, C.R.; Remsen, E.E. 1998. Emulsion polymerization process for organically soluble and electrically conducting polyaniline. *Macromolecules* 31:1735–1744.
3. Watkins, J.D.; Marken, F. 2016. Application of ionic liquids, emulsions, sonication, and microwave assistance. In *Organic Electrochemistry*, 5th ed. Hamerich, O.; Speiser, B. (eds.) CRC Press, London, p. 336.
4. Pletcher, D.; Walsh, F.C. 1993. *Industrial Electrochemistry*. Chapman & Hall, London, p. 298.
5. Feess, H.; Wendt, H. 1982. Performance of electrolysis with two-phase electrolyte. In *Techniques of Electro-Organic Synthesis*. Part III, Weinberg, N.L.; Tilak, B.V. (eds.) John Wiley & Sons, New York, p. 81.
6. Mackay, R.A.; Texter, J. 1982. *Electrochemistry in Colloids and Dispersions*. Wiley-VCH, Weinheim, Germany.
7. Menger, F.M.; Rhee, J.U.; Rhee, H.K. 1975. Applications of surfactants in synthetic organic chemistry. *J. Org. Chem.* 40:3803–3805.
8. Francke, R.; Little, R.D. 2014. Redox catalysis in organic electrosynthesis: Basic principles and recent developments. *Chem. Soc. Rev.* 43:2492–2521.
9. Eberson, L.; Helgee, B. 1975. Studies on electrolytic substitution reactions. IX. Anodic cyanation of aromatic ethers and amines in emulsions with the aid of phase transfer agents. *Acta Chem. Scand.* 29B:451–456.
10. Lee, G.A.; Freedman, H.H. 1976. Phase transfer catalyzed oxidations of alcohols and amines by aqueous hypochlorite. *Tetrahedron Lett.* 20:1641–1644.
11. Pletcher, D.; Tomov, N. 1977. The use of electrogenerated hypobromite for the phase transfer catalysed oxidation of benzyl alcohols. *J. Appl. Electrochem.* 7:501–504.
12. Chou, T.C.; Cheng, C.H. 1992. Indirect anodic oxidation of anthracene to anthraquinone in a slurry electrolyte systems in the presence of both surfactant and redox mediator. *J. Appl. Electrochem.* 22:743–748.
13. Ellis, S.R.; Pletcher, D.; Brooks, W.N.; Healy, K.P. 1983. Electrosynthesis in systems of two immiscible liquids and a phase transfer catalyst. V. The anodic chlorination of naphthalene. *J. Appl. Electrochem.* 13:735–741.
14. Forsyth, S.R.; Pletcher, D.; Healy, K.P. 1987. Electrosynthesis in systems of two immiscible liquids and a phase transfer catalyst. VII. The chlorination of substituted naphthalenes. *J. Appl. Electrochem.* 17:905–913.
15. Ellis, S.R.; Pletcher, D.; Gough, P.; Korn, S.R. 1982. Electrosynthesis in systems of two immiscible liquids and a phase transfer catalyst. I. The anodic cyanation of naphthalene. *J. Appl. Electrochem.* 12:687–691.
16. Ellis, S.R.; Pletcher, D.; Gamlen, P.H.; Healy, K.P. 1982. Electrosynthesis in systems of two immiscible liquids and a phase transfer catalyst. II. Aromatic nuclear acyloxylation. *J. Appl. Electrochem.* 12:693–699.
17. Laurent, E.; Rauniyar, G.; Thomalla, M. 1984. Anodic substitutions in emulsions under phase transfer catalysis conditions. I. Cyanation of dimethoxybenzenes. *J. Appl. Electrochem.* 14:741–748.
18. Knarr, R.F.; Velasco, M.; Lynn, S.; Tobias, C.W. 1992. The electrochemical reduction of 2-ethyl anthraquinone. *J. Electrochem. Soc.* 139:948–954.

19. Cognet, P.; Berlan, J.; Lacoste, G.; Fabre, P.L.; Jud, J.M. 1996. Application of metallic foams in an electrochemical pulsed flow reactor. 2. Oxidation of benzyl alcohol. *J. Appl. Electrochem.* 26:631–637.

20. Floner, D.; Laglaine, L.; Moinet, C. 1997. Indirect electrolysis involving an ex-cell two- phase process. Reduction of nitrobenzenes with a titanium complex as mediator. *Electrochim. Acta* 42:525–529.

21. Cortona, M.N.; Vettorazzi, N.R.; Silber, J.J.; Sereno, L.E. 1999. Electrochemical nitration of naphthalene in the presence of nitrite ion in aqueous non-ionic surfactant solutions. *J. Electroanal. Chem.* 470:157–165.

22. Huissoud, A.; Tissot, P. 1999. Electrochemical reduction of 2-ethyl-9,10-anthraquinone (EAQ) and mediated formation of hydrogen peroxide in a two-phase medium—Part I: Electrochemical behaviour of EAQ on a vitreous carbon rotating disc electrode (RDE) in the two-phase medium. *J. Appl. Electrochem.* 29:11–16.

23. Tanaka, H.; Kuroboshi, M.; Takeda, H.; Kanda, H.; Torii, S. 2001. Electrochemical asymmetric epoxidation of olefins by using an optically active Mn-salen complex. *J. Electroanal. Chem.* 507:75–81.

24. Jan, T.; Dupas, B.; Floner, D.; Moinet, C. 2002. Electrosynthesis of dibenzonaphthyridine derivatives from 2,2-(2-nitrobenzyl)-2-substituted-acetonitriles. *Tetrahedron Lett.* 43:5949–5952.

25. Raju, T.; Kalpana Devi, G.; Kulangiappar, K. 2006. Regioselective bromination of toluene by electrochemical method. *Electrochim. Acta* 51:4596–4600.

26. Sripriya, R.; Chandrasekaran, M.; Noel, M. 2008. Electrochemical homolytic and heterolytic coupling of activated olefins in the absence and presence of benzyl bromide in microemulsion. *J. Appl. Electrochem.* 38:597–603.

27. Watkins, J.D.; Ahn, S.D.; Taylor, J.E.; Bull, S.D.; Bulman-Page, P.C.; Marken, F. Liquid–liquid electro-organo-synthetic processes in a carbon nanofibre membrane microreactor: Triple phase boundary effects in the absence of intentionally added electrolyte. *Electrochim. Acta* 56:6764–6770.

28. Christopher, C.; Lawrence, S.; Bosco, A.J.; Xavier, N.; Raja, S. 2012. Selective oxidation of benzyl alcohol by two phase electrolysis using nitrate as mediator. *Catal. Sci. Tech.* 2:824–827.

29. Zhao, R. 2012. Electrosynthesis of sodium hypochlorite in room temperature ionic liquids and in situ electrochemical epoxidation of olefins. *React. Kinet. Mech. Catal.* 106:37–47.

30. Li, C.; Zeng, C.-C.; Hu, L.M.; Yang, F.-L.; Joon Yoo, S.; Little, R.D. 2013. Electrochemically induced C-H functionalization using bromide ion/2,2,6,6-tetramethylpiperidinyl-N-oxyl dual redox catalysts in a two-phase electrolytic system. *Electrochim. Acta* 114:560–566.

31. Jing, R.; Tang, J.; Zhang, Q.; Chen, L.; Ji, D.X.; Yu, F.W.; Lu, M.Z.; Ji, J.B.; Wang, J.L. 2015. An insight into the intensification of aqueous/organic phase reaction by the addition of magnetic polymer nanoparticles. *Chem. Eng. J.* 280:265–274.

32. Anelli, P.L.; Biffi, C.; Montanari, F.; Quici, S. 1987. Fast and selective oxidation of primary alcohols to aldehydes or to carboxylic-acids and of secondary alcohols to ketones mediated by oxoammonium salts under 2-phase conditions. *J. Org. Chem.* 52:2559–2562.

33. Tang, J.; Zhang, Q.; Hu, K.C.; Zhang, P.; Wang, J.L. 2017. Novel high TEMPO loading magneto-polymeric nanohybrids: An efficient and recyclable Pickering interfacial catalyst. *J. Catal.* 353:192–198.

34. Rusling, J.F. 1991. Controlling electrochemical catalysis with surfactant microstructures. *Acc. Chem. Res.* 24:75–81.

35. Rusling, J.F.; Zhou, D.L. 1997. Electrochemical catalysis in microemulsions. Dynamics and organic synthesis. *J. Electroanal. Chem.* 439:89–96.
36. Budnikova1, Y.H. 2002. Metal complex catalysis in organic electrosynthesis. *Uspekhi Khimi* 71:126–158.
37. Zhou, D.L.; Carrero, H.; Rusling, J.F. 1996. Radical vs anionic pathway in mediated electrochemical reduction of benzyl bromide in a bicontinuous microemulsion. *Langmuir* 12:3067–3074.
38. Cheng, W.; Compton, R.G. 2016. Quantifying the electrocatalytic turnover of vitamin B_{12} mediated dehalogenation on single soft nanoparticles. *Angew. Chem. Int. Ed.* 55:2545–2549.
39. Large, M.J.; Ogilvie, S.P.; Meloni, M.; Graf, A.A.; Fratta, G.; Salvage, J.; King, A.A.K.; Dalton, A.B. 2018. Functional liquid structures by emulsification of graphene and other two-dimensional nanomaterials. *Nanoscale* 10:1582–1586.
40. Texter, J. 2015. Graphene oxide and graphene flakes as stabilizers and dispersing aids. *Curr. Opin. Colloid Interface Sci.* 20:454–464.
41. Anastas, P.T.; Warner, J.C. 1998. *Green Chemistry: Theory and Practice.* Oxford University Press, New York, p. 30.
42. Wadhawan, J.D.; Marken, F.; Compton, R.G. 2001. Biphasic sonoelectrosynthesis. A review. *Pure Appl. Chem.* 73:1947–1955.
43. Banks, C.E.; Davies, T.J.; Evans, R.G.; Hignett, G.; Wain, A.J.; Lawrence, N.S.; Wadhawan, J.D.; Marken, F.; Compton, R.G. 2003. Electrochemistry of immobilised redox droplets: Concepts and applications. *Phys. Chem. Chem. Phys.* 5:4053–4069.
44. Gulaboski, R.; Riedl, K.; Scholz, F. 2003. Standard Gibbs energies of transfer of halogenate and pseudohalogenate ions, halogen substituted acetates, and cycloalkyl carboxylate anions at the water backslash nitrobenzene interface. *Phys. Chem. Chem. Phys.* 5:1284–1289.
45. Li, Y.; Deng, H.Q.; Dick, J.E.; Bard, A.J. 2015. Analyzing benzene and cyclohexane emulsion droplet collisions on ultramicroelectrodes. *Anal. Chem.* 87:11013–11021.
46. Raynal, F.; Barhdadi, R.; Perichon, J.; Savall, A.; Troupel, M. 2002. Water as solvent for nickel-2,2′-bipyridine-catalysed electrosynthesis of biaryls from haloaryls. *Adv. Synth. Catal.* 344:45–49.
47. Watkins, J.D.; Amemiya, F.; Atobe, M.; Bulman-Page, P.C.; Marken, F. 2010. Liquid | liquid biphasic electrochemistry in ultra-turrax dispersed acetonitrile | aqueous electrolyte systems. *Electrochim. Acta* 55:8808–8814.
48. Banks, C.E.; Rees, N.V.; Compton, R.G. 2002. Sonoelectrochemistry in acoustically emulsified media. *J. Electroanal. Chem.* 535:41–47.
49. Davies, T.J.; Banks, C.E.; Nuthakki, B.; Rusling, J.F.; France, R.R.; Wadhawana, J.D.; Compton, R.G. 2002. Surfactant-free emulsion electrosynthesis via power ultrasound: electrocatalytic formation of carbon–carbon bonds. *Green Chem.* 4:570–577.
50. Atobe, M.; Ikari, S.; Nakabayashi, K.; Amemiya, F.; Fuchigami, T. 2010. Electrochemical reaction of water-insoluble organic droplets in aqueous electrolytes using acoustic emulsification. *Langmuir* 26:9111–9115.
51. Nakabayashi, K.; Fuchigami, T.; Atobe, M. 2013. Tandem acoustic emulsion, an effective tool for the electrosynthesis of highly transparent and conductive polymer films. *Electrochim. Acta* 110:593–598.
52. Wadhawan, J.D.; Del Campo, F.J.; Compton, R.G.; Foord, J.S.; Marken, F.; Bull, S.D.; Davies, S.G.; Walton, D.J.; Ryley, S. 2001. Emulsion electrosynthesis in the presence of power ultrasound biphasic Kolbe coupling processes at platinum and boron-doped diamond electrodes. *J. Electroanal. Chem.* 507:135–143.

53. Marken, F.; Watkins, J.D.; Collins, A.M. 2011. Ion-transfer- and photo-electrochemistry at liquid I liquid I solid electrode triple phase boundary junctions: Perspectives. *Phys. Chem. Chem. Phys.* 13:10036–10047.

54. MacDonald, S.M.; Watkins, J.D.; Bull, S.D.; Davies, I.R.; Gu, Y.; Yunus, K.; Fisher, A.C.; Page, P.C.B.; Chan, Y.; Elliott, C.; Marken, F. 2009. Two-phase flow electrosynthesis: Comparing N-octyl-2-pyrrolidone-aqueous and acetonitrile-aqueous three-phase boundary reactions. *J. Phys. Org. Chem.* 22:52–58.

55. MacDonald, S.M.; Watkins, J.D.; Gu, Y.; Yunus, K.; Fisher, A.C.; Shul, G.; Opallo, M.; Marken, F. 2007. Electrochemical processes at a flowing organic solvent I aqueous electrolyte phase boundary. *Electrochem. Commun.* 9:2105–2110.

56. Gosh, J.C.; Bhattacharyya, S.K.; Muthanna, M.S.; Mitra, C.R. 1952. Electrolytic reactions on porous carbon anodes: I. The preparation of p-benzoquione by the oxidation of benzene. *J. Sci. Ind. Res. (India)*, 11B:356–360.

57. Gosh, J.C.; Bhattacharyya, S.K.; Rao, M.R.A.; Muthanna, M.S.; Patnaik, P.B. 1952. Electrolytic reactions on porous carbon anodes: II. The preparation of chlorobenzene from benzene. *J. Sci. Ind. Res. (India)* 11B:361–364.

58. Fenton, J.M.; Alkire, R.C. 1988. Mass-transport in flow-through porous-electrodes with 2-phase liquid-liquid flow. *J. Electrochem. Soc.* 135:2200–2209.

59. Watkins, J.D.; Ahn, S.D.; Taylor, J.E.; Bull, S.D.; Bulman-Page, P.C.; Marken, F. 2011. Liquid-liquid electro-organo-synthetic processes in a carbon nanofibre membrane microreactor: Triple phase boundary effects in the absence of intentionally added electrolyte. *Electrochim. Acta* 56:6764–6770.

60. Watkins, J.D.; Taylor, J.E.; Bull, S.D.; Marken, F. 2012. Mechanistic aspects of aldehyde and imine electro-reduction in a liquid-liquid carbon nanofiber membrane microreactor. *Tetrahedron Lett.* 53:3357–3360.

61. Watkins, J.D.; Lawrence, R.; Taylor, J.E.; Bull, S.D.; Nelson, G.W.; Foord, J.S.; Wolverson, D.; Rassaei, L.; Evans, N.D.M.; Gascon, S.A.; Marken, F. 2010. Carbon nanoparticle surface functionalisation: Converting negatively charged sulfonate to positively charged sulfonamide. *Phys. Chem. Chem. Phys.* 12:4872–4878.

62. del Campo, F.J. 2014. Miniaturization of electrochemical flow devices: A mini-review. *Electrochem. Commun.* 45:91–94.

63. Green, R.A.; Brown, R.C.D.; Pletcher, D. 2015. Understanding the performance of a microfluidic electrolysis cell for routine organic electrosynthesis. *J. Flow Chem.* 5:31–36.

64. Green, R.A.; Brown, R.C.D.; Pletcher, D. 2016. Electrosynthesis in extended channel length microfluidic electrolysis cells. *J. Flow Chem.* 6:191–197.

65. Scialdone, O.; Galia, A.; Sabatino, S.; Mira, D.; Amatore, C. 2015. Electrochemical conversion of dichloroacetic acid to chloroacetic acid in a microfluidic stack and in a series of microfluidic reactors. *Chem. Electro. Chem.* 2:684–690.

66. Wouters, B.; Hereijgers, J.; De Malsche, W.; Breugelmans, T.; Hubin, A. 2016. Electrochemical characterisation of a microfluidic reactor for cogeneration of chemicals and electricity. *Electrochim. Acta* 210:337–345.

67. Zimmerman, W.B. 2011. Electrochemical microfluidics. *Chem. Eng. Sci.* 66:1412–1425.

68. Gu, Y.F.; Fisher, A.C. 2013. An ac voltammetry approach for the detection of droplets in microfluidic devices. *Analyst* 138:4448–4452.

69. Yunus, K.; Marks, C.B.; Fisher, A.C.; Allsopp, D.W.E.; Ryan, T.J.; Dryfe, R.A.W.; Hill, S.S.; Roberts, E.P.L.; Brennan, C.M. 2002. Hydrodynamic voltammetry in microreactors: multiphase flow. *Electrochem. Commun.* 4:579–583.

70. Kaluza, D.; Adamiak, W.; Kalwarczyk, T.; Sozanski, K.; Opallo, M.; Jonsson-Niedziolka, M. 2013. Anomalous effect of flow rate on the electrochemical behavior at a liquid | liquid interface under microfluidic conditions. *Langmuir* 29:16034–16039.
71. Kaluza, D.; Adamiak, W.; Opallo, M.; Jonsson-Niedziolka, M. 2014. Comparison of ion transfer thermodynamics at microfluidic and droplet-based three phase electrodes. *Electrochim. Acta* 132:158–164.

8 Solid-to-Solid Transformations in Organic Electrosynthesis

Antonio Doménech-Carbó
University of Valencia

CONTENTS

8.1 Introduction 167
8.2 Antecedents: Solid-State Electrochemistry 167
8.3 Solid-to-Solid Electrochemical Transformations 168
8.4 Solid-to-Solid Electrosynthesis of Organic Compounds........................... 170
8.5 Catalytic Electrosynthesis 173
8.6 Concluding Remarks 176
References.. 176

8.1 INTRODUCTION

There is a wide tradition in the organic electrosynthesis occurring in solution phase, i.e., when the parent compounds and the resulting components are all species in solution in a suitable electrolyte. Although much less extended, there is the possibility, however, of producing organic electrosynthesis via solid-to-solid interconversion of compounds immobilized onto electrode surfaces. This chapter is devoted to present the foundations, problems, and future developments of this electrosynthetic scenario in the context of solid-state electrochemistry, taken as that referring to electrochemical systems where at least one solid ionic conductor displays redox processes [1,2]. Here, the focus will be electrosynthetic processes involving microparticulate deposits of solids on inert electrodes, avoiding other related methodologies, in particular those involving polymer coatings on electrodes, with obvious direct relation but having different specific aspects.

8.2 ANTECEDENTS: SOLID-STATE ELECTROCHEMISTRY

Taken in the aforementioned meaning, contemporary solid-state electrochemistry was initiated in the 1960s by the studies on solids embedded into carbon paste electrodes due to Adams, Kuwana, and coworkers [3,4] and extended by Bauer, Brainina, Zakharchuk, and others in the subsequent decade [5–7]. For electrosynthetic purposes, however, the essential issue was the development of the voltammetry of immobilized particles (VIMP) methodology by Scholz and coworkers in the late 1980s [8–11]. The VIMP consists of the record of the voltammetric response of a set of solid particles abrasively attached to an inert electrode (typically paraffin-impregnated graphite) which is put in contact with an electrolyte in which the solid is insoluble or sparingly soluble. As described in comprehensive reviews [12–14], this methodology has been applied to a variety of materials constituting a basis for the electrochemical analysis of solids because the obtained voltammetric response is representative of the chemical composition and structure of the solid, being influenced by the granulometry of the sample and, of course, by the nature of the electrolyte [15,16].

8.3   SOLID-TO-SOLID ELECTROCHEMICAL TRANSFORMATIONS

The electrochemical processes taking place in the VIMP can be roughly divided into solid-state transformations and reductive/oxidative dissolutions. Although the latter can also be of synthetic interest, the focus will be made on the former. In turn, these can be divided in electrochemical processes involving or not phase segregation, i.e., when the solid electrochemically generated occupies the same or a different phase than the parent solid.

Generically, electrochemical processes involving microparticulate deposits on inert electrodes can be described, as illustrated in Figure 8.1, in terms of electron transfer between the base electrode and the solid particles necessarily coupled, by reasons of charge conservation, to ion transfer between the particles and the electrolyte. According to the theoretical modeling developed by Lovric, Scholz, Oldham, and coworkers [17–21], the electrochemical reaction for a ion-permeable solid is initiated at the three-phase junction between the base electrode, the solid particle, and the electrolyte and proceeds via electron transfer across the electrode/particle interface and ion transfer between the electrolyte/particle interface. The propagation

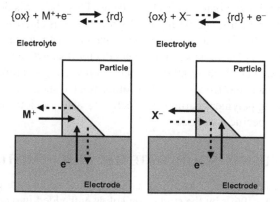

FIGURE 8.1   Schematic representation of the reduction (solid arrows) and oxidation (dotted arrows) of a solid particle immobilized onto the surface of an inert electrode in the cases in which the solid was permeable to cations (M^+) and anions (X^-).

of the redox reaction through the solid particle involves electron hopping between immobile redox centers coupled to ion diffusion through available interstitial positions in the solid.

In this scheme, symmetric situations appear for the cases in which the solid was permeable to cations or to anions. As schematized in Figure 8.1, in the first case, reduction (oxidation) involves the ingress (issue) of cations, and in the second case, reduction (oxidation) involves the issue (ingress) of anions in/to the solid. Such processes can be represented, assuming reversibility, by means of the equations:

$$\{ox\}_{solid} + M^{+}_{solv} + e^{-} \rightleftarrows \{rd\cdots M^{+}\}_{solid} \tag{8.1}$$

$$\{rd\}_{solid} + X^{-}_{solv} \rightleftarrows \{ox\cdots X^{-}\}_{solid} + e^{-} \tag{8.2}$$

In the inorganic domain, the typical example of this kind of electrochemistry was that of metal hexacyanoferrates, typically Prussian blue [22–24], with many other examples [12–16]. In the organic field, however, the currently studied solid-state electrochemistry is dominated by redox processes in contact with aqueous electrolytes involving proton transfer coupled to electron transport, formally able to being theoretically described on the basis of the previously mentioned modeling [17–21], in terms of coupled electron and proton diffusion [25]. This involves, rather than occupying interstitial positions in the solid lattice, as generally occurring in inorganic solids, the formation/destruction of covalent bonds in the solid. The electrochemical reduction and oxidation of indigo in the presence of aqueous electrolytes can be considered as a paradigmatic case [26]. As schematized in Figure 8.2, solid indigo can either be reduced, via two-proton, two-electron process, to solid leucoindigo and oxidized, via another two-proton, two-electron process, to dehydroindigo.

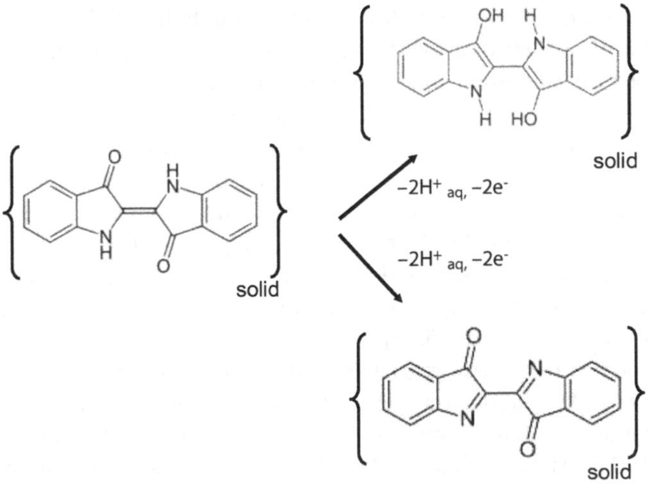

FIGURE 8.2 Scheme of redox processes in the solid-state electrochemistry of indigo in contact with aqueous electrolytes.

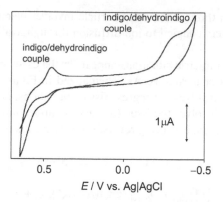

FIGURE 8.3 Cyclic voltammogram of indigo microparticulate deposit on paraffin-impregnated graphite electrode immersed into 0.25 M aqueous sodium acetate buffer at pH 4.75. Potential scan rate is 20 mV s^{-1}.

Interestingly, both processes exhibit significant electrochemical reversibility (see Figure 8.3) [26,27]. This has to be underlined because both the oxidation and reduction processes involve a significant reorganization of chemical bonds in the parent molecule and, additionally, significant modifications of the intermolecular hydrogen bonding, a feature particularly relevant in the solid state. In general, however, reversibility is uneasily attained in solid-state electrochemistry of organic solids [28,29].

8.4   SOLID-TO-SOLID ELECTROSYNTHESIS OF ORGANIC COMPOUNDS

In principle, the advance of the redox process through the solid particles may occur via topotactic transformation of the parent solid into its product of oxidation or reduction forming initially a solid solution of the second on the former. The solid product, however, can segregate from the original solid, then appearing as inclusions in the parent compound [30] or forming bilayered structures [31,32]. The existence of miscibility gaps has been theoretically modeled by Lovric et al. [33]. The case of 7,7,8,8-tetracyanoquinodimethane in contact with aqueous electrolytes has been studied by Bond et al. [34–36]. Upon redox cycling, nanocrystals of this compound yield solid–solid phase transformations controlled by nucleation and growth of the new crystalline phase [34], often giving rise to new solids with vast differences in the morphology and crystal size [35,36].

Figure 8.4 illustrates the morphological changes observed, via atomic force microscopy (AFM) imaging, on morin crystals attached to a graphite plate submitted to an oxidative potential input in contact with aqueous phosphate buffer at pH 7.0 using AFM-VIMP experiments [37,38]. Morin, such as other flavonoids, is quasi-reversibly oxidized in contact with aqueous media, this electrochemistry involving in much cases a solid-state catechol to quinone transformation [39,40]. The progress

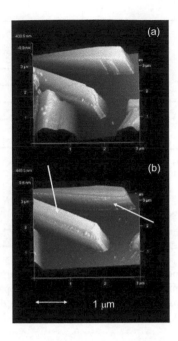

FIGURE 8.4 Topographic AFM image of a set of morin crystals deposited onto a graphite plate in contact with 0.10 M aqueous potassium phosphate buffer at pH 7.0, (a) before and (b) after applying a potential input of 0.65 V vs. Ag/AgCl for 2 min. (Courtesy of M.T. Doménech-Carbó, Polytechnical University of Valencia.)

of the reaction is denoted by the modification of the crystals surface, with an apparent delamination (marked by arrows in Figure 8.4) in principle corresponding to the replacement of the parent morin by its oxidized form.

For electrosynthetic purposes, it is pertinent to remark that there are different possible situations. Figure 8.5 depicts idealized representations, the cases in which the daughter compound forms: (a) a solid solution within the parent compound, (b) immiscible inclusions within the parent compound, and (c,d) a layer segregated from the parent compound. Here, two possibilities have been drawn: when (c) the diffusivity of protons is larger than that of electrons ($D_{H^+} > D_e$), and (d) when the proton diffusivity is lower than the electron diffusivity ($D_{H^+} < D_e$). The crystal pattern, however, can adopt a variety of forms. An example can be seen in Figure 8.6, corresponding to the oxidative attack on a crystalline deposit of clonazepam, a benzodiazepine displaying a well-defined solid-state electrochemistry [41], on a graphite plate in contact with 0.10 M aqueous sodium acetate buffer at pH 4.75. After applying a potential input of 0.45 V vs. Ag/AgCl for 5 min, the cluster of fine acicular crystals become apparently enlarged transversally and partially delaminated. This can be interpreted in terms of an individualized advance of the reaction layer through the different crystal bars of clonazepam.

In several reported cases, parent and/or intermediate species, often adsorbed, in solution are involved in the generation of organic solids. This is the case of the

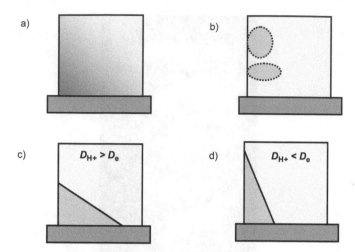

FIGURE 8.5 Idealized representations of several cases of organic electrosynthesis where the daughter compound forms: (a) a solid solution within the parent compound; (b) immiscible inclusions within the parent compound; a layer segregated from the parent compound when (c) $D_{H^+} > D_e$, and (d) $D_{H^+} < D_e$.

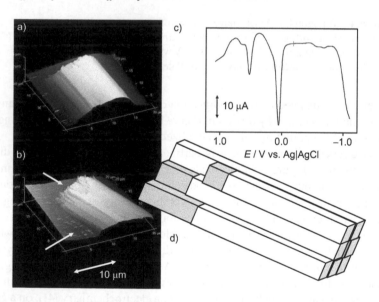

FIGURE 8.6 Topographic AFM image of a crystalline deposit of clonazepam on a graphite plate in contact with 0.10 M aqueous sodium acetate buffer at pH 4.75, (a) before and (b) after applying a potential input of 0.45 V vs. Ag/AgCl for 5 min. (c) Square wave voltammogram of a clonazepam-modified graphite electrode in contact with the above electrolyte; potential scan initiated at −1.05 V vs. Ag/AgCl in the positive direction; potential step increment 4 mV; square wave amplitude 25 mV; frequency 5 Hz. (d) Pictorial representation of the presumed redox reaction advance. (AFM images courtesy of M.T. Doménech-Carbó, Polytechnical University of Valencia; voltammogram from Doménech-Carbó, A. et al., *J. Solid State Electrochem.*, 14, 465, 2010. With permission.)

oxidation of 2-mercaptobenzoxazole to bis(benzoxazolyl) disulfide [42] and azobenzene [43]. In the case of diphenylamine, studied by Inzelt [44], solid crystals mechanically attached to platinum and gold electrodes produced oxidative dimerizations and polymerizations in contact with nonaqueous and/or mixed solvent solutions subsequently resulting in reversible transformations between aromatic amine and quinoidal imine forms. Conversely, solid-phase organic synthesis associated to polystyrene [45,46] and polythiophene [47] polymers with electrochemically cleavable groups has been reported.

8.5 CATALYTIC ELECTROSYNTHESIS

Apart from the above "direct" solid-to-solid electrosynthetic methodologies, there is the possibility of using different electrocatalytic systems for synthetic purposes. In this context, Marken and coworkers have explored new strategies involving redox mediators acting catalytically on solid reagents [48–51]. The mediation of 8-dihydroxyanthraquinone in the reduction of colloidal indigo [48], although yielding leucoindigo in solution, can be viewed as a strategy usable for solid-to-solid electrosynthesis. The reaction of water-insoluble microcrystalline solids with redox mediators in solution phase is illustrated by the solid-to-solid conversion of diphenylcarbinol into benzophenone mediated by several 2,2,6,6-tetramethyl-1-piperidinyloxy (TEMPO) radicals [49,50]. The overall reaction was found to follow an EC' (electro-chemical catalytic) mechanism, where nucleation and growth of the product was followed by a diffusive step termed as "pore transport" where benzophenone forms inclusions into the crystals of the diphenylcarbinol [51]. Remarkably, the reaction is initiated by the one-electron oxidation of TEMPO to its cationic form, a well-established electrochemical process [52], whereas both the substrate and the reaction product are electroinactive.

From a structural point of view, it is interesting to mention that the advance of solid-to-solid electrochemical reactions can follow structural patterns differing from those in Figure 8.5. This would be the case of the TEMPO-mediated solid-state diphenylcarbinol/benzophenone reaction for which a pictorial scheme illustrating the proposed initial nucleation and growth stage and the pore transport stage is depicted in Figure 8.7 [51].

The above scheme is structurally in agreement with the formation of pore systems in organometallic crystals submitted to solid-state electrochemistry reductions/oxidations. Representative examples can be seen in Figure 8.8, where electron microscopy images for the reduction of Cu(MOF) accompanied by insertion of alkaline cations in aqueous media [53,54] and the oxidation of $[\{Au_3Cu_2(C_2C_6H_4Fc)_6\}$ $Au_3(PPh_2C_6H_4PPh_2)_3]$ $(PF_6)_2$, an Au(I)–Cu(I) heterotrimetallic cluster bearing ferrocenyl groups, accompanied by insertion of chloride anions in contact with aqueous NaCl solution [55], are depicted. In the case of Cu(MOF), the original crystals forming microparticulate deposits on graphite plates were submitted to a reductive potential input yielding an external deposit of Cu metal. As a result, the crystals become apparently crossed by channels denoting that the electrochemical reaction advanced through determined crystallographic planes and/or structural domains. In the case of the crystals of heterotrimetallic cluster, the oxidative step, consisting of

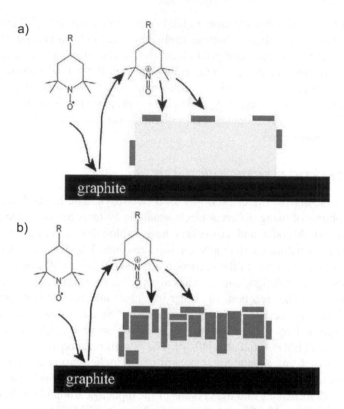

FIGURE 8.7 TEMPO-assisted solid-to-solid conversion of diphenylcarbinol into benzophenone electrochemically driven. Scheme of (a) the initial nucleation and growth and (b) the pore transport stages of the reaction. (Reproduced from Kaluza, D. et al., *J. Solid State Electrochem.*, 19, 1277, 2015. With permission.)

the oxidation of ferrocenyl units coupled to chloride insertion, results in the apparent decapping and formation of holes which can also be interpreted in terms of the formation of inclusions of the Cl⁻-containing oxidized phase.

Heterogenization of the catalyst is another electrosynthetic strategy. This is the case of Kolbe C–C electrosynthesis using piperidine supported on silica gel [56], and carbohydrate oxidation with 4-benzoyloxy-TEMPO heterogenized in a microporous polymer [50]. As participating in the VIMP methodology, one can mention possible electrosynthetic processes associated to organic substrates and catalysts attached to inorganic nanoporous supports. Catalysis of 1,4.dihydrobenzoquinone electrochemical oxidation by 4,4'-bipyridinium ions attached to silica and mesoporous aluminosilica supports [57] and light-driven *N,N,N',N'*,-tetramethylbenzidine oxidation by 6-nitro-1',3,3'-trimethylspiro[2*H*-1-benzopyran-2,2'-indoline] anchored to zeolite Y and MCM-41 silicates [58] would be examples of this type of methodology. This last case, whose electrocatalytic effect is illustrated in Figure 8.9, incorporates a photoelectrocatalytic effect, which is of potential application for synthetic purposes.

FIGURE 8.8 Transmission electron microscope images recorded for Cu/MOF crystals in contact with 0.50 M sodium acetate buffer at pH 4.85, (a) before and (b) after application of a constant potential of −1.0 V during 10 min. Scanning electron microscope images of a deposit of crystals of [{Au$_3$Cu$_2$(C$_2$C$_6$H$_4$Fc)$_6$}Au$_3$(PPh$_2$C$_6$H$_4$PPh$_2$)$_3$](PF$_6$)$_2$: (c) before and (d) after application of an oxidative potential step at 0.75 V for 1 h in contact with 0.10 M NaCl. (From Doménech-Carbó, A. et al., *Electrochem. Commun.*, 8, 1830, 2006 (a,b); Doménech-Carbó, A. et al., *J. Phys. Chem. C*, 111, 13701, 2007 (c,d). With permission.)

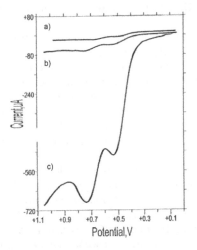

FIGURE 8.9 Linear potential scan voltammograms for 3 mM solution of *N,N,N',N'*,- tetramethylbenzidine in 0.10 M Bu$_4$NPF$_6$/MeCN at (a) unmodified paraffin-impregnated graphite electrode, and that electrode modified with 6-nitro-1',3',3'-trimethylspiro[2*H*-1-benzopyran-2,2'-indoline] anchored to (b) MCM-41 silicate and (c) zeolite Y. Potential scan rate 100 mV s^{-1}. (From Doménech-Carbó, A. et al., *J. Phys. Chem. B*, 108, 20064, 2004. With permission.)

8.6 CONCLUDING REMARKS

Solid-to-solid electrosynthesis can be regarded as a research field currently being in its early stages. It seems obvious that, in the current stage of development, solid-to-solid electrochemically driven reactions offer some of the capabilities of solution-phase electrosynthesis: the possibility of control of experimental conditions and the absence of lateral reactions, in particular, but also several drawbacks, in particular, the difficulty in the isolation of the products of the reaction. Other aspects, such as the incorporation of electrolyte counterions and the peculiar structural features often appearing in the involved solid-to-solid interconversion processes, can be considered as ambivalent, as far as the difficult control of the synthesized structures is balanced by the opening of possibilities similar to those of the well-known cationic/anionic doping in conducting polymers. In this regard, the possibility of in situ synthesizing of solid catalysts for electrochemical processes (electrosynthetic or electroanalytical) gives considerable potential interest to this methodology. Future research lines should incorporate investigation of mechanisms of electrochemical processes, elucidation/control of structural features, implementation of different experimental materials (solvents, base electrodes, supporting electrolytes), arrangements (different modes of particle immobilization, use of particle dispersions), and analytical strategies (use of redox mediators, use of photoelectrochemical catalysis) in order to expand the scope of solid-to-solid electrosynthetic methods.

REFERENCES

1. Bard, A.J.; Inzelt, G.; Scholz, F. Eds. 2008. *Electrochemical Dictionary*. Berlin-Heidelberg: Springer.
2. Scholz, F. 1997. Welcome to all readers and authors of the journal of solid state electrochemistry. *J. Solid State Electrochem.* 1:1.
3. Adams, R.N. 1958. Carbon paste electrodes. *Anal. Chem.* 30:1576–1576.
4. Kuwana, T.; French, W.G. 1964. Electrooxidation or reduction of organic compounds into aqueous solutions using carbon paste electrode. *Anal. Chem.* 36:241–242.
5. Bauer, D.; Gaillochet, M.P. 1974. Etude du comportement de la pate de carbone a compose electroactif incorporé. *Electrochim. Acta* 19:597–606.
6. Brainina, Kh.Z.; Vidrevich, M.B. 1981. Stripping analysis of solids. *J. Electroanal. Chem.* 121: 1–28.
7. Belyi, V.I.; Smirnova, T.P.; Zakharchuk, N.F. 1989. Phase composition and structure of native oxides on $A^{III}B^{V}$ semiconductors. *Appl. Surf. Sci.* 39:161–167.
8. Scholz, F.; Nitschke, L.; Henrion, G. 1989. A new procedure for fast electrochemical analysis of solid materials. *Naturwissenschaften* 76:71–72.
9. Scholz, F.; Nitschke, L.; Henrion, G.; Damaschun, F. 1989. A new technique to study the electrochemistry of minerals. *Naturwissenschaften* 76:167–168.
10. Scholz, F.; Nitschke, L.; Henrion, G. 1989. Identification of solid materials with a new electrochemical technique—the abrasive stripping voltammetry. *Fresenius' Zeitschrift Anal. Chem.* 334:56–58.
11. Scholz, F.; Nitschke, L.; Henrion, G.; Damaschun, F. 1989. Abrasive stripping voltammetry—the electrochemical spectroscopy for solid state: application for mineral analysis. *Fresenius' Zeitschrift Anal. Chem.* 335:189–194.

12. Scholz, F.; Meyer, B. 1994. Electrochemical solid state analysis—state of the art. *Chem. Soc. Rev.* 23:341–347.
13. Scholz, B.; Meyer, B. 1998. Voltammetry of solid microparticles immobilized on electrode surfaces, in *Electroanalytical Chemistry, A Series of Advances*, Bard, A.J.; Rubinstein, I. Eds., 20, pp. 1–86. Marcel Dekker, New York.
14. Grygar, T.; Marken, F.; Schröder, U.; Scholz, F. 2002. Voltammetry of microparticles: a review. *Coll. Czech. Chem. Commun.* 67:163–208.
15. Doménech-Carbó, A.; Labuda, J.; Scholz, F. 2013. Electroanalytical chemistry for the analysis of solids: characterization and classification. *Pure Appl. Chem.* 85:609–631.
16. Scholz, F.; Schröder, U.; Gulaboski, R.; Doménech-Carbó, A. 2014. *Electrochemistry of Immobilized Particles and Droplets*, 2nd ed. Springer, Berlin-Heidelberg.
17. Lovric, M.; Scholz, F. 1997. A model for the propagation of a redox reaction through microcrystals. *J. Solid State Electrochem.* 1:108–113.
18. Oldham, K.B. 1998. Voltammetry at a three-phase junction. *J. Solid State Electrochem.* 2:367–377.
19. Lovric, M.; Hermes, M.; Scholz, F. 1998. The effect of the electrolyte concentration in the solution on the voltammetric response of insertion electrodes. *J. Solid State Electrochem.* 2:401–404.
20. Lovric, M.; Scholz, F. 1999. A model for the coupled transport of ions and electrons in redox conductive microcrystals. *J. Solid State Electrochem.* 3:172–175.
21. Schröder, U.; Oldham, K.B.; Myland, J.C.; Mahon, P.J.; Scholz, F. 2000. Modelling of solid state voltammetry of immobilized microcrystals assuming an initiation of the electrochemical reaction at a three-phase junction. *J. Solid State Electrochem.* 4:314–324.
22. Dostal, A.; Meyer, B.; Scholz, F.; Schroder, U.; Bond, A.M.; Marken, F.; Shaw, S.J. 1995. Electrochemical study of microcrystalline solid Prussian blue particles mechanically attached to graphite and gold electrodes: electrochemically induced lattice reconstruction. *J. Phys. Chem.* 99:2096–2103.
23. Scholz, F.; Dostal, A. The formal potentials of solid metal hexacyanometalates. *Angew. Chem. Int. Ed.* 34:2685–2687.
24. Kahlert, H.; Retter, U.; Lohse, H.; Siegler, K.; Scholz, F. 1998. On the determination of the diffusion coefficients of electrons and of potassium ions in copper(II) hexacyanoferrate(II) composite electrodes. *J. Phys. Chem. B* 102:8757–8765.
25. Doménech-Carbó, A.; Doménech-Carbó, M.T. 2006. Chronoamperometric study of proton transfer/electron transfer in solid state electrochemistry of organic dyes. *J. Solid State Electrochem.* 10:949–958.
26. Bond, A.M.; Marken, F.; Hill, E.; Compton, R.G.; Hügel, H. 1997. The electrochemical reduction of indigo dissolved in organic solvents and as a solid mechanically attached to a basal plane pyrolytic graphite electrode immersed in aqueous electrolyte solution. *J. Chem. Soc., Perkin Trans.* 2.2:1735–1742.
27. Doménech-Carbó, A.; Martini, M.; de Carvalho, L.M.; Doménech-Carbó, M.T. 2012. Square wave voltammetric determination of the redox state of a reversibly oxidized/reduced depolarizer in solution and in solid state. *J. Electroanal. Chem.* 684:13–19.
28. Doménech-Carbó, A.; Navarro, P.; Arán, V.J.; Muro, B.; Montoya, N.; García-España, E. 2010. Selective electrochemical discrimination between dopamine and phenethylamine derived psychotropic drugs using electrodes modified with an acyclic receptor containing two terminal 3-alkoxy-5-nitroindazole rings. *Analyst* 135:1449–1455.
29. Imperatore, C.; Persico, M.; Aiello, A.; Luciano, P.; Guiso, M.; Sanasi, M.F.; Taramelli, D.; Parapini, S.; Cebrián-Torrejón, G.; Doménech-Carbó, A.; Fattorusso, C.; Menna, M. 2015. Marine inspired antiplasmodial thiazinoquinones: synthesis, computational studies and electrochemical assays. *RSC Adv.* 5:61006–61011.

30. Hermes, M.; Lovric, M.; Hartl, M.; Retter, U.; Scholz, F. 2001. On the electrochemically driven formation of bilayered systems of solid Prussian-blue-type metal hexacyanoferrates: a model for Prussian blue | cadmium hexacyanoferrate supported by finite difference simulations. *J. Electroanal. Chem.* 501:193–204.

31. Pickup, P.G.; Leidner, C.R.; Denisevich, P.; Willman, K.W.; Murray, R.W. 1984. Bilayer electrodes: theory and experiment for electron trapping reactions at the interface between two redox polymer films. *J. Electroanal. Chem.* 164:39–61.

32. Leidner, C.P.; Denisevich, P.; Willman, K.W.; Murray, R.W. 1984. Charge trapping reactions in bilayer electrodes. *J. Electroanal. Chem.* 164:63–78.

33. Lovric, M.; Hermes, M.; Scholz, F. 2000. Solid state electrochemical reactions in systems with miscibility gaps. *J. Solid State Electrochem.* 4:394–401.

34. Bond, A.M.; Fletcher, S.; Marken, F.; Shaw, S.J.; Symons, P.J. 1996. Electrochemical and X-ray diffraction study of the redox cycling of nanocrystals of 7,7,8,8- tetracyanoquinodimethane. Observation of a solid–solid phase transformation controlled by nucleation and growth. *J. Chem. Soc., Faraday Trans.* 92:3925–3933.

35. Nafady, A.; Al-Qahtani, N.J.; Al-Farhan, K.A.; Bhargava, S.; Bond, A.M. 2014. Synthesis and characterization of microstructured sheets of semiconducting $Ca[TCNQ]_2$ via redox-driven solid-solid phase transformation of TCNQ microcrystals. *J. Solid State Electrochem.* 18:851–859.

36. Nafady, A.; Sabri, Y.M.; Kandjani, A.E.; Alsalmel, A.M.; Bond, A.M.; Bhargava, S. 2016. Preferential synthesis of highly conducting Tl(TCNQ) phase II nanorod networks via electrochemically driven TCNQ/Tl(TCNQ) solid-solid phase transformation. *J. Solid State Electrochem.* 20:3303–3314.

37. Doménech-Carbó, A.; Doménech-Carbó, M.T. 2008. In situ AFM study of proton-assisted electrochemical oxidation/reduction of microparticles of organic dyes. *Electrochem. Commun.* 10:1238–1241.

38. Doménech-Carbó, A.; Doménech-Carbó, M.T.; Calisti, M.; Maiolo, V. 2010. *J. Solid State Electrochem.* 14:465–477.

39. Doménech-Carbó, A.; Doménech-Carbó, M.T.; Saurí-Peris, M.C. 2005. Electrochemical identification of flavonoid dyes in work of art samples by abrasive voltammetry at paraffin- impregnated graphite electrodes. *Talanta* 66:769–782.

40. Doménech-Carbó, A.; Doménech-Carbó, M.T.; Calisti, M.; Maiolo, V. 2010. Sequential identification of organic dyes using the voltammetry of microparticles approach. *Talanta* 81:404–411.

41. Doménech-Carbó, A.; Martini, M.; de Carvalho, L.M.; Viana, C.; Doménech-Carbó, M.T.; Silva, M. 2013. Screening of pharmacologic adulterant classes in herbal formulations using voltammetry of microparticles. *J. Pharmaceut. Biomed. Anal.* 74:194–204.

42. Schaufβ, A.; Wittstock, G. 1999. Oxidation of 2-mercaptobenzoxazole in aqueous solution: solid phase formation at glassy carbon electrodes. *J. Solid State Electrochem.* 3:361–369.

43. Komorsky-Lovric, S. 1997. Voltammetry of azobenzene microcrystals. *J. Solid State Electrochem.* 1:94–99.

44. Inzelt, G. 2002. Cyclic voltammetry of solid diphenylamine crystals immobilized on an electrode surface and in the presence of an aqueous solution. *J. Solid State Electrochem.* 6:265–271.

45. Meldal, M.; Auzanneau, F.I.; Hindsgaul, O.; Palcic, M.M. 1994. A PEGA resin for use in solid phase chemical–enzymatic synthesis of glycopeptides. *J. Chem. Soc., Chem. Commun.* 1849–1850.

46. Schuster, M.; Wang, P.; Paulson, J.C.; Wong, C.H. 1994. Solid-phase chemical-enzymic synthesis of glycopeptides and oligosaccharides. *J. Am. Chem. Soc.* 116:1135–1136.

47. Dubey, S.; Fabre, B.; Marchand, G.; Pilard, J.F.; Simonet, J. 1999. Voltammetric investigation of new polythiophene derivatives possessing electrochemically cleavable arylsulfonamide groups as precursors for solid phase electrosynthesis. *J. Electroanal. Chem.* 477:121–129.

48. Vuorema, A.; John, P.; Toby, A.; Jenkins, A.; Marken, F. 2006. A rotating disc voltammetry study of the 1,8-dihydroxyanthraquinone mediated reduction of colloidal indigo. *J. Solid State Electrochem.* 10:865–871.

49. Jin, Y.; Edler, K.J.; Marken, F.; Scout, J.L. 2014. Voltammetric optimisation of TEMPO-mediated oxidations at cellulose fabric. *Green Chem.* 16:3322–3327.

50. Kolodziej, A.; Ahn, S.D.; Carta, M.; Malpass-Evans, R.; McKeown, N.B.; Chapman, R.S.L.; Bull, S.D.; Marken, F. 2015. Electrocatalytic carbohydrate oxidation with 4-benzoyloxy-TEMPO heterogenised in a polymer of intrinsic microporosity. *Electrochim. Acta* 160:195–201.

51. Kaluza, D.; Jöhnsson-Niedziólka, M.; Ahn, S.D.; Owen, R.E.; Jones, M.D.; Marken, F. 2015. Solid-solid EC' TEMPO-electrocatalytic conversion of diphenylcarbinol to benzophenone. *J. Solid State Electrochem.* 19:1277–1283.

52. Liaigre, D.; Breton, T.; Beigsir, E.M. 2005. Kinetic and selectivity control of TEMPO electro-mediated oxidation of alcohols. *Electrochem. Commun.* 7:312–316.

53. Doménech-Carbó, A.; García, H.; Doménech-Carbó, M.T.; Llabrés-i-Xamena, F. 2006. Electrochemical nanometric patterning of MOF particles: anisotropic metal electrodeposition in Cu/MOF. *Electrochem. Commun.* 8:1830–1834.

54. Doménech-Carbó, A.; García, H.; Doménech-Carbó, M.T.; Llabrés-i-Xamena, F. 2007. Electrochemistry of metal-organic frameworks: a description from the voltammetry of microparticles approach. *J. Phys. Chem. C* 111:13701–13711.

55. Doménech-Carbó, A.; Koshevoy, I.O.; Montoya, N.; Pakkanen, T.A. 2010. *Electrochem. Commun.* 12:206–209.

56. Kurihara, H.; Fuchigami, T.; Tajima, T. 2008. Kolbe carbon-carbon coupling electrosynthesis using solid-supported bases. *J. Org. Chem.* 73:6888–6890.

57. Doménech-Carbó, A.; Álvaro, M.; Ferrer, B.; García, H. 2003. Electrochemistry of mesoporous organosilica of MCM-41 type containing 4,4'-bipyridinium units: voltammetric response and electrocatalytic effect on 1,4-dihydrobenzoquinone oxidation. *J. Phys. Chem. B* 107:12781–12788.

58. Doménech-Carbó, A.; García, H.; Casades, I.; Esplá, M. 2004. Electrochemistry of 6-nitro-1',3,3'-trimethylspiro[2H-1-benzopyran-2,2'-indoline] associated to zeolite Y and MCM-41 aluminosilicate. Site-selective electrocatalytic effect on N,N,N',N'-tetramehtylbenzidine oxidation. *J. Phys. Chem. B* 108:20064–20075.

Index

A

Acrylonitrile, 90
Acyliminium, 3
Alkoxylation, 103
Alloy, 112
Amination, 105
Ammonia, 57
Aniline, 53
Anodic processes, 24
Aromatic amidation, 28
Atomic force microscopy, 170
ATRP, 141
Au-Cu-alloy, 113

B

Benzaldehyde, 83
Biphasic, 149
Birch reduction, 22
Boron-doped diamond, 97
Bucky paper, 159

C

C, C-cleavage, 105
Carbon dioxide reduction, 99, 113, 14
Carbon nanofibers, 159
Carboxylation, 23
Carnot thermal cycle, 46
Cascade reaction, 30
Catalytic electrosynthesis, 173
Cathodic processes, 22, 30
Cation pool, 3
Cavitation, 86
C-H activation, 24
Channel flow, 161, 8
Chiral electrodes, 2
Click chemistry, 27
Cobalamin, 69
Cogeneration, 45
Computer simulation, 10
Conducting polymer, 128
Copolymer, 130
Couette flow, 9
Cyanation, 135, 104

D

Deep eutectic solvents, 13
Dehydrogenation, 56

DFT, 9
Diels-Alder, 20, 32
Diffusion layer thickness, 85
Diffusion layer, 8
Dimerization, 21
Direct transformation, 15
Dispersion, 89
Divided, 7
DMFC, 54
DMSO, 17

E

Electride, 116
Electroactive polymer, 140
Electroorganic reaction, 3
Electropolymerization, 129, 91
Emulsion, 149

F

Ferrocene, 29
Fick's law, 8
Flow electrolysis, 4
Fluorination, 136, 104
Fluorodesulfurization, 137
Fuel cell, 46
Functional polymer, 127

G

Galvanostatic, 6
Graphite, 18

H

Heck reaction, 30
Heterogenization, 174
Hexafluoroisopropanol, 108, 27
H-type divided cell, 83, 3
Hydrodynamic methods, 8
Hydrogen peroxide, 54
Hydroxyl radical, 98

I

Imine, 160
Immersion horn, 83
Indigo, 170
Intramolecular cyclization, 73
Ionic liquids, 11

K

Kolbe reactioin, 157, 87, 15

L

Leaded bronze, 22
Light-assisted elecrolysis, 72

M

Mass transfer, 85
Mass transport, 7
Mediator system, 68
Membrane electrode assembly (MEA), 53
Methane, 58
Methoxylation, 19
Microfluidic, 161, 23
Microjetting, 86
Microreactor, 4
Microreactor, 8, 16
Molten salt, 62
Multiphase, 89

N

Nitroarene, 102
Nitrobenzene, 53

O

Oximes, 101
Ozone, 98

P

P3MT, 133
Paired reaction, 5, 32, 137
Paired transformations, 31
Pb cathode, 101, 110
Pd catalyst, 26, 30
PEMFC, 52
Phenol coupling, 107
Phenol-arene cross-coupling, 109
Photoelectrochemistry, 67
Photoelectrosynthesis, 67
Photo-energy, 67, 16
Photovoltaic device, 76
Polyaniline, 139
Poly-ethylene-glycol, 9
Polymer solvents, 14
Polypyrrole, 91

Polysilane, 131
Post-functionalization, 133
Potentiostatic, 6
Propionitrile, 90
Prussian blue, 169

R

Radical, 74
Reaction layer, 8
Redox mediator, 150, 24, 30
Rocking disk, 8
Rotating disk, 8
RTILs, 11–13

S

Sacrificial Anode, 22
Salt bridge, 7
Self-supported, 4
Semiconductor, 18
Shunt path, 69
Slurry electrode, 89
Sn-Pb-alloy, 114
Solar cell, 77
Solid oxide fuel cell (SOFC), 57
Solid state electrochemistry, 167
Solvated electron, 2, 23
Solvents, 10
Sonoelectrosynthesis, 81
Sonotrode, 84
Sulfonamides, 29
Surfactant, 154

T

TEMPO, 173, 154
Three-phase junction, 168
Triple phase boundary, 157

U

Ultrasound, 155, 81
Ultra-turrax, 156
Undivided, 7, 29

V

Vinylsulfones, 21
Vitamin B12, 155, 69
Voltammetry of immobilised particles, 168

Milton Keynes UK
Ingram Content Group UK Ltd.
UKHW040055071024
449327UK00019B/586

9 780367 732875